W9-BYD-615

ELECTRICAL ENGINEERING DESIGN COMPENDIUM

ELECTRICAL ENGINEERING DESIGN COMPENDIUM

Robert L. McConnell
Wils L. Cooley
West Virginia University

Nigel T. Middleton
Colorado School of Mines

ADDISON-WESLEY PUBLISHING COMPANY
Reading, Massachusetts • Menlo Park, California
New York • Don Mills, Ontario • Wokingham, England
Amsterdam • Bonn • Sydney • Singapore • Tokyo
Madrid • San Juan • Milan • Paris

Library of Congress Cataloging-in-Publication Data

McConnell, Robert L.
   Electrical engineering design compendium / Robert L. McConnell,
Wils L. Cooley, Nigel T. Middleton ; sponsored by the National
Science Foundation.
      p.   cm.
   ISBN 0–201–56612–5
   1. Electronic circuit design.   I. Cooley, Wils L.   II. Middleton,
Nigel T., 1954– .   III. Title.
TK7867.M375 1993
621.3815--dc 20   92-27714
                                                                   CIP

2 3 4 5 6 7 8 9 10-BA-959493

Preface

Instructors of traditional undergraduate electrical engineering courses can use the exercises in this compendium to increase the design content in their courses without having to expend major effort, reorganize the courses, or displace current materials. The compendium is intended to supplement conventional electrical engineering texts rather than be a stand-alone design book. Ideally, students would obtain the book as they begin their studies and instructors in electrical engineering courses would assign its exercises in lieu of some of the usual drill and analysis problems. Because analysis is an integral part of the design process, no loss in the practice of analysis should occur by instructors' doing this. The exercises extend into practical design settings the theoretical and analytical concepts covered in conventional texts and the content covers the major disciplines within the field.

We have covered a broad range of topics in each major discipline of electrical engineering; however, we recognize it is impossible to cover them all. Similarly, rather than attempting to cover all aspects of design, we focus on those that are appropriate in engineering science and analysis courses. Therefore this book will find its greatest utility in the basic courses of the electrical engineering curriculum.

Most treatises on design and most design project courses tend to focus on global and system-level processes and decisions, those that relate to such topics as market opportunity, performance features, and political and social interaction. Such complex issues often require much effort and time to resolve and cannot easily be integrated into a theory and analysis course without displacing some of the current material. We leave the system and higher-level issues to projects and courses dedicated to design.

Similarly, we do not address intermediate, or system-level issues such as technology, functional operation, and the manufacturing process. Instead, we put our attention on the nitty-gritty details of circuit and component-level design. The design exercises in this book illustrate practical applications of the material covered in engineering science and analysis-oriented courses. Students usually must be proficient with design at the circuit and component level before they are ready to carry out a system-level design, such as that required for a senior design project.

We intentionally selected exercises with limited scope so students could complete the designs in no more than a few hours and so the instructor can easily evaluate them. In many cases, the exercise statement specifies a circuit configuration. The design process then consists of selecting and specifying appropriate real-world components that will provide the desired circuit performance. We feel this approach is realistic because entry-level designers often work in such environments. We also believe that students should not be expected to invent novel circuit configurations or reinvent a particular circuit configuration until their design experience has matured. However, students should gain valuable experience by completing these exercises and in the process, develop a repertoire of circuit configurations upon which to draw for more open-ended design exercises.

ORGANIZATION OF THE BOOK

The book is divided into three parts, one each for tutorials, data, and exercises. The tutorials cover design methodologies a student might need to know to complete many of the exercises. We include this material because electrical engineering curricula seldom do. Chapter 1 briefly describes the working environment of a design engineer. Chapter 2 discusses the design process and how to present a design for evaluation. Chapter 3 deals with design methods that consider component value variations resulting from manufacturing tolerances or arising from temperature change. It demonstrates how to develop a robust design, that is, one in which a finished device will meet specifications, even with the variations (within tolerance) of the specified components as they are selected at random during mass production. Chapter 4 discusses reliability in general and how to predict the reliability of a circuit, a topic essential to competitive product design. Chapter 5 deals with safety, specifically, the Failure Modes and Effects Analysis (FMEA), another topic seldom discussed formally but of considerable importance for responsible design. Chapter 6 is a short discussion on design optimization. It talks about how to balance conflicting design criteria such as cost, performance, and reliability, and outlines procedures for optimizing the design.

The tutorials are intended only to introduce the topics discussed. The student can acquire from them an adequate working knowledge needed for the elementary applications in the exercises. Bibliographies are provided at the end of each chapter for those students wishing to explore further.

Part 2 contains what can be considered a list of **approved** parts. Lists of approved parts are common in many companies and designers at these companies usually have to select design components from those lists. Also included in this section are specifications of the approved parts and cost and reliability information for use in those exercises in which economy and reliability are design criteria. For most exercises, students should use the approved parts listed in Part 2; however, in some exercises, special parts are specified.

Part 3 exercises are organized into the traditional topics within electrical engineering. However, many exercises are appropriate for design activities in more than one topic area. Each exercise illustrates a technical concept and many specify a design method or criterion. We invite the instructor to modify any exercise to better correspond to the objectives in a particular course.

We hope that inclusion of the design exercises into present electrical engineering courses will begin to return design practice to its rightful place in the engineering curriculum. We are convinced that improving the design abilities of our engineers can help our industries become more competitive in the international marketplace.

ACKNOWLEDGMENTS

This manuscript was developed under USE-8854664 grant from the National Science Foundation, Division of Undergraduate Science, Engineering, and Mathematics Education, in the Education and Human Resources Directorate. The design exercises derive from many sources. Some were developed by the book's authors for use in the courses they have taught and so bear no acknowledgment. Those suggested or contributed by other engineers or educators carry acknowledgments. Individuals of particular note who contributed many exercises include Bill Graff and Paul Leiffer of LeTourneau University, Paul Gray of the University of Wisconsin-Platteville, Keith Stanek of the University of Missouri-Rolla, and Manos Roumeliotis of City College of Thessaloniki. To the extent that their exercises are sound, all of our contributors deserve full credit. However, any errors are the responsibility of the authors alone, as these contributors have not had the opportunity to proofread their submissions.

We wish to acknowledge our colleagues at West Virginia University; Dr. Edward Ernst, formerly with the National Science Foundation and now with the University of South Carolina; and a host of other engineers and engineering educators throughout the nation for their advice and encouragement during the development of the compendium. We received much help from David Thomas in collecting materials and Dennis Yost in organizing and editing the manuscript as well as preparing the reliability data. Bonna Musick, who typed much of the manuscript, and Maggie LeMasters were a great help in taking care of a host of administrative problems and generally helping to keep the chaos under some degree of control. The acknowledgments would not be complete, however, without mentioning the support and patience of our wives, Debby, Jane, and Glynis. To all, we thank you.

Introduction to Students

We developed this compendium because we realized that many of our students were having difficulty applying the basic theory and analytic methods discussed in their electrical engineering courses to the design tasks in their senior design course. A typical design problem can have many acceptable solutions, thus requiring the designer to make decisions both in circuit configuration and in component selection. However, our students weren't accustomed to working in an unstructured environment where there is no single correct solution; consequently, the prospect of making decisions in a real design situation caused anxiety, immobilizing some students. Others reverted to former unskilled methods of solving problems. They ignored or forgot the analytic techniques learned in class and tried to design using an experimental process, resulting in circuits littered with potentiometers that they adjusted until an acceptable circuit condition arose. These breakdowns were not altogether unexpected, for few students had any design experience requiring them to integrate theory, analysis, and use of real-world components.

We wanted to give you a chance to create designs that are closely related to the theoretical topic being studied and to do so while the topic is still fresh in your mind, not weeks later in a seemingly unrelated or loosely related laboratory exercise. Therefore, we selected design exercises we feel illustrate the application of theoretical principles and analytic methods at various points in basic electrical engineering courses. Ideally, your instructor will assign a design exercise at an appropriate time with the expectation that you design the circuit or system using the theory and methods recently covered in the course.

Some exercises require a functional design in which the circuit or system is expected to work or provide a proper function when nominal component values are used. In other cases, the situation calls for designs that will meet more rigorous criteria. In these instances, you would be expected to design the circuit or system as though it is intended for mass production and therefore must operate within specification in spite of component tolerances. In still other cases, you will be asked to design to meet extended criteria such as reliability, safety, optimum performance, or minimum cost. The tutorials in Part 1 should help you develop these more demanding designs.

Part 2 is a collection of approved components and their specifications. Unless otherwise indicated, you will be expected to design using approved components. In this age of integrated circuit technology, many system designs called for in the exercises can be implemented with a single medium- or large-scale integrated circuit and in the real world, a design using the IC would be realistic. However, to provide the desired design experience and to illustrate pedagogical principles, you will be designing with discrete components and small-scale integrated circuits in the exercises.

Practicing engineers design in many types of working environments, so we used a variety of such environments in the exercises in Part 3. The application's setting often is given to help you understand the motivation for the requirements and to assist your making the necessary judgments given that background. You will find that some exercise settings are whimsical, while others represent more realistic situations.

We hope you enjoy these exercises and find the design experience interesting and rewarding. Best of luck!

The Authors

Contents

ELECTRICAL ENGINEERING DESIGN COMPENDIUM

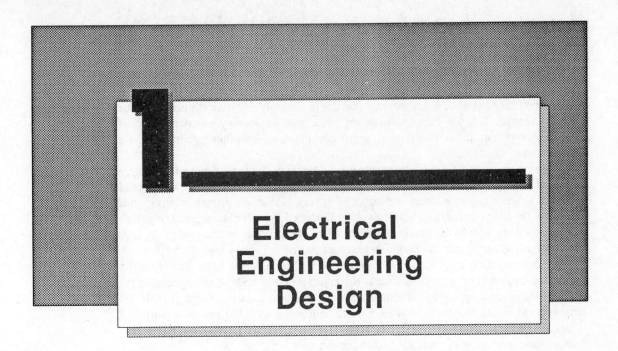

1

Electrical Engineering Design

"Engineering is the creative art of applying science for the benefit of humankind."

J. H. Wujek,
Apple Computers, Inc.
1989

1.1 INTRODUCTION

We live in a technological age surrounded by devices and systems not even imagined a century ago; many were unheard of only a generation ago. We take for granted our vast highway network and extensive electric power distribution systems. We have long accepted intricate transportation systems that encompass the ground, air, and water and sophisticated communication systems whose signals are carried by wire, radio waves, optical fibers, and orbiting satellites. We have seen computer technology grow from the first electronic computers in the 1930s to present-day personal computers and supercomputers. Our lives continue to be enriched by these and many other technological innovations. Although many different individuals and companies have contributed the concepts, inventions, and money necessary to advance technology, it is engineers who have provided the technical expertise to make it happen. And as an electrical engineering student, you will soon have the opportunity to contribute your particular talents to a profession which is providing exciting new possibilities for society.

1.2 WHAT DOES AN ENGINEER DO?

Because engineering talent is required at many stages in the developmental process of technology, many different but nevertheless related engineering functions are involved. In a large organization, the functions of each individual engineering group might be quite distinct, while in a small organization the divisions might be blurred, but the same functions are still carried out. In either case, it is common that an engineering design group comprise engineers from multiple disciplines. In the following discussion, we describe the engineering functions carried out during the development and life of a product.

In industry, the process of developing and producing a new product usually begins in one of two ways. Someone might recognize a potential new product that could satisfy a perceived need or a new method that could possibly improve the performance of an existing product. Or, the process might result from a company's desire to exploit a particular invention or component or to complement its other products. Let's consider an example.

Suppose we have an idea for an electronic automotive carburetor that will improve fuel economy by 25%. What would it take to get the new carburetor into widespread use? We would need to address a number of issues. Some are global, dealing with such considerations as fuel availability and fuel safety, if new fuel blends are required; meeting EPA standards for pollution; disposal of the product at the end of its life; customer acceptance; and the economic impact on the company and its employees. Others, at the system level for the automobile, could include the car's fuel handling system; its engine size, if performance is different; its body style; engine metallurgy; and the manufacturing process. Because we don't intend to focus on these global- and system-level issues and because we are more concerned with the types of work and issues faced by a young engineer just starting a career, let's narrow our focus and follow the engineering process for just the carburetor.

First, we would need to develop a prototype to demonstrate the carburetor's feasibility. The job of doing this would be assigned to a **development group** that would comprise engineers with the appropriate expertise in electronics and gasoline engine technology and that would be concerned primarily with getting the device to operate properly. As it worked to solve the problems associated with merging these two technologies, this group might create several models before devising one that worked correctly.

Once the carburetor's feasibility had been demonstrated and the decision made to produce the device, another group of engineers, the **design group,** would be called on to refine it, that is, to make it inexpensive to manufacture and reliable and efficient to operate. This design group, which might include all or part of the development group, would focus on designing improved circuits and components. At this stage, there would be considerable emphasis on the importance of minimizing the cost and maximizing the reliability of the carburetor; because of the large volume of product that would be built, pennies per unit saved in manufacture or repair could save the company tens of thousands of dollars yearly. Next, to gather statistical performance data, a preproduction batch of carburetors would be fleet-tested and a detailed market survey and economic analysis would be conducted. Also, a comprehensive set of workshop manuals for carburetor tuning and repair would be written.

When the carburetor is released for manufacturing, the **manufacturing group,** which might contain engineers from several disciplines, would track the manufacturing process. As the engineers gained experience with the process, they would make further refinements, such as in the product design or in the manufacturing process itself. Then, as the carburetors came off the production line, a **quality control group** would be responsible for monitoring the quality of the devices. And finally, once the carburetor is in widespread use, more engineers would be involved in tracking problems that might arise and in further testing and developing the product.

If this product were a large piece of equipment or a business machine, **field** or **customer service engineers** would be involved to ensure it is used correctly and in a manner most beneficial to the customer. They also would track the utility and effectiveness of the device as it is used by customers and notify the manufacturing group of possible design modifications that could improve later versions of the item. Further, as they helped customers solve problems associated with their using the device, the engineers could tell those customers about related products that might be helpful.

In this discussion, we have focused on the process of manufacturing a consumer product. However, this process is similar in utility industries and in those industries that produce business or defense products or one-of-a-kind items; the same engineering functions are performed, only the personnel and organization are different.

We have also laid out the steps of the engineering process as if they occurred sequentially. This was often the case in the past, but global pressure to reduce time delays in engineering has changed the picture. Much engineering is now done **concurrently**, meaning that a multidisciplinary team carries out the steps all together, making field support and manufacturing decisions along with technology development decisions.

As we have seen, throughout the life of a product, from conception to application, engineers apply technology and their scientific knowledge with the aim of solving problems and satisfying needs. At any level, as engineers work on a product, they are designing in one form or another; at any step, an individual engineer might contribute creative ideas that could improve the product or enhance its usefulness. In addition, engineers must be alert to and design to meet the challenge of keeping costs down while achieving reliability, safety, and optimum performance.

1.3 TOP-DOWN DESIGN

Product development progresses through a three-level process known as *top-down design*. In the beginning stages, design decisions are made at a **global level.** These decisions include determining which features will satisfy the perceived need and provide the necessary interactions with other equipment or with operators. The design effort next moves to the **system level,** which involves resolving technological issues, such as producing functional block diagrams to show how the product will operate. Finally, at the **circuit or device level,** the details of the design are filled in; among other actions, the functional blocks are designed and dimensions, components, etc., are specified. There are technological and creative challenges unique to each level. However, this progression is not necessarily linear. At any time, the engineer might need to return to an earlier stage as additional evaluation reveals a flaw in or alternatives to earlier decisions (see Fig. 1.1). Note that although the process involves returning to the top level from the bottom level, it always goes through the system level on the way down.

In a typical manufacturing company, the three stages of design are handled by engineers of different experience levels. Global design usually is the responsibility of engineers and managers who are experienced in political, social, and economic concerns. System design generally is handled by project engineers who often are the project leaders during the developmental phase. Circuit and device design usually is carried out by recently hired engineers who are less experienced but nonetheless familiar with the latest technology.

We focus the remaining discussion and all the exercises in this compendium on circuit and device-level design. By doing this, we don't intend to imply that the other levels of design are not important; rather, we want to stress that in order to be an effective designer at the system and global levels, the electrical engineer first must achieve proficiency at the circuit and device level.

1.4 DESIGN ACTIVITIES

The engineer as designer is faced with a multitude of tasks and concerns, such as the following:

■ Analyze the problem to determine the characteristics the circuit must have.

- Choose a circuit that has those characteristics, or invent one that does.
- Determine the operational and functional limits of the circuit.
- Select components that will provide proper operation.
- Verify by analysis that the circuit will function as required.

Throughout this process, the designer must not only keep in mind the functional requirements of the circuit under design but also be aware of the system requirements. The design must simultaneously satisfy both of these.

Further, the designer must consider the circuit's safety, reliability, and cost. The system must operate safely. Its reliability and the cost of both manufacturing and maintenance affect its competitiveness. Component failure cannot be allowed to endanger the operation of other parts of the system or to become a safety hazard to the equipment operator or to others. Loss of the system is annoying to the customer and could result in considerable inconvenience and expense for loss of service and repairs. Failure might even cause injury or death.

As a designer, you will have these concerns. The quality of the product will depend on your designing skill and ability. Thus it's not only in your best interests to design as well as you can; your fellow employees will come to depend on you to contribute to the continued prosperity of your organization. It follows also that because many industries compete in a global market, the economic well-being of your community may be affected as well; many dramatic shifts in corporate power and prosperity among the developed and developing nations have been driven by quality product design.

1.5 ABOUT THIS COMPENDIUM

Including design exercises in your courses is intended to improve your design skill by exposing you to several design criteria, methods, and situations. In addition, you also will obtain a better understanding of the different specialties of electrical engineering.

The exercises in this compendium cover the gamut of electrical engineering, always focusing on circuit and device level issues. Some exercises are intended to direct your attention toward creating circuit configurations; others deal with designing circuits that can be consistently manufactured despite normal component variations; still others focus on reliability and safety.

Figure 1.1. The Top-Down Design Process.

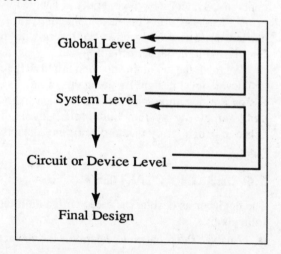

Because these issues are seldom covered in lecture or laboratory classes in the curriculum, Part 1—Chapters 1–6—presents tutorials that discuss several aspects of design. Chapter 1 has outlined the design process, and Chapter 2 describes how to present your design for evaluation. The remaining tutorial chapters discuss the following four issues that must be considered in design:

1. Designing for mass production (Chapter 3)
2. Reliability prediction (Chapter 4)
3. Designing for safety (Chapter 5)
4. Design optimization (Chapter 6)

Many of the methods and techniques outlined in the tutorials are the same as those used in industry. We recommend that before you complete the first design assignment, you finish reading Chapter 1 and also read Chapter 2. If you choose, you can defer Chapters 3–6 until such time as a design assignment requires you know their contents.

Part 2 of the book is the data section. The components listed are **approved** parts and are to be used in all designs unless you are instructed otherwise. Many companies maintain lists of approved parts that they buy in large quantities and over which they are willing to maintain quality control. While this section is far from complete, it is representative of available parts. We have included the names and specifications of several different types of resistors and capacitors as well as several semiconductor devices, integrated circuits, and other hardware. Giving you the specifications will enable you to learn how to read specification sheets and to gain experience in designing with real components in such a way that the final designs are ready for manufacturing.

The bulk of the book contains the design exercises. Each exercise is intended to take only a few hours to complete and many require the methods and techniques given in the tutorials. Sometimes, a circuit configuration is given in the exercise statement; in these cases you are to specify the components to be used in the circuit construction. Other times, you are left to decide on a configuration. These latter cases are more open-ended and often result in widely varying solutions. Both situations are encountered in industry by engineers.

You are encouraged to use the resources in this book to learn effective design techniques and develop good designs. We are confident that in the process of your doing the design exercises, you will become a better engineer.

2

The Process and Presentation of Engineering Design

"... a piece of machinery, or anything that is made, is like a book, if you can read it."

Henry Ford

2.1 INTRODUCTION

As part of their varied professional responsibilities, engineers are *designers,* the intellectual craftsmen of new technological products and systems needed by industry and society. As designers, they play two roles in the development of a new product or system. First, they synthesize solutions to problems by creating a plan, or *design,* that consists of an orderly composition of technical and practical issues for implementing the product or system. Second, they document and present the final design for evaluation and implementation.

In this chapter, we discuss these two roles as they relate to the design of electrical or electronic devices and circuits. Before proceeding, however, let's clarify the definitions of devices and circuits. A *device* is a component in a circuit. A *circuit* is an electrically connected network of devices. The specifications of a device, including its nominal value, depend on the dimensions and physical characteristics of the materials used in the device and on the configurations of those materials within the device and of the device itself. Examples of electrical and electronic devices are resistors, capacitors, inductors and transformers, semiconductors, battery cells, antennas, and motors. Circuits are described by the configuration of the interconnections between the components, or devices, and the **value** of each component within the network.

2.2 THE DESIGN PROCESS

Designing devices or circuits involves the following four steps:

1. Set the design objectives.
   This is done primarily to establish the design's performance requirements.

7

2. Select the device or circuit configuration.
3. Select the configuration parameters, that is, the material types and dimensions for a device or the component values for a circuit.
4. Predict the performance.

This is done by analyzing the combination of the selected configuration and the parameters to compare the predicted performance of the design with the performance requirements. When a selected combination meets these requirements, then those performance predictions become the performance specifications of the final design.

The designer might need to revisit these steps several times before reaching a result that best meets the objectives. In some cases, the configuration of the device or circuit is given at the outset of the design process, so step 2 could be bypassed. In other cases, multiple configurations might be possible, so the designer might select the parameters (step 3) for each configuration and analyze its performance (step 4) before settling on one particular configuration. Or, if the analysis called for in step 4 shows that the performance requirements are not satisfied, the designer then would need to return to one of the earlier steps to select either a different configuration or different parameter values. Let's consider an example.

Example 2.1. An LED Indicator.

An LED (light emitting diode) is to be used as a power-on indicator for a piece of equipment we're building. We have run experiments on the LED and found that for the indicator to be bright enough, we need to deliver 25 mA or more through the LED. We also learned that the voltage across the LED was 2.1 V over the current range of 25–30 mA.

Design a circuit that will deliver this current to the LED when powered from a 5.0-V dc supply.

Step 1: Set the Design Objectives.

We need to design a circuit that will deliver at least 25 mA to a LED whose voltage drop is 2.1 V (see Fig. 2.1). Because more current than necessary is wasteful, the design should minimize the current but still stay above 25 mA.

We should be as specific as possible in setting these objectives and recognize that there are three categories of statements that can be made. The first category concerns the **performance requirements** of the design. These requirements quantify those

---

Figure 2.1. Diagram of LED Indicator Circuit.

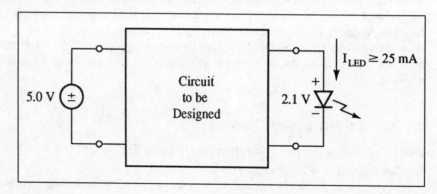

features the design must exhibit at the very minimum. The second category itemizes the **design constraints** that embody those parameters or configuration features of the design that are set by factors typically not controlled by a designer. During the design process, we have no flexibility to change any of these constrained features. On the other hand, we have complete freedom of choice in matters beyond these constraints, although our choices should be influenced by the third category of statements, which involve the **design criteria**. In seeking a final design that is influenced by these design criteria, we will be generating a good design, that is, a design that not only meets the performance requirements, but does so effectively.

Therefore we can state the following design objectives regarding our example:

Performance Requirements:	$I_{LED} \geq 25\ mA$
Design Constraints:	$V_{SUPPLY} = 5.0\ V$
	$V_{LED} = 2.1\ V$
Design Criteria:	The excess of I_{LED} over 25 mA should be a minimum.

Step 2: Select the Configuration.

Many different circuit configurations, several of which are quite complex, could perform the function of limiting the LED current to about 25 mA. However, the simplest circuit that will do the job is usually the best choice, for reasons that include high reliability and low cost, so let's try the simplest circuit configuration we can imagine. In this case, such a configuration would use a series resistor to limit the current, as shown in Fig. 2.2.

Step 3: Select the Configuration Parameters.

Next, we need to decide which resistor value to use. This step requires the specific knowledge and analysis techniques of an electrical engineer. To start with, we set a target of 25 mA through the LED. Then we apply Kirchhoff's voltage law to the circuit to establish that the resistor will have 2.9 V across it. Because it has 25 mA through it, we can use Ohm's law to calculate the required resistance, as follows:

$$R = \frac{2.9}{0.025}\ \Omega \tag{2.1}$$

giving

$$R = 116\ \Omega.$$

Figure 2.2. Proposed Circuit Configuration for Example 2.1.

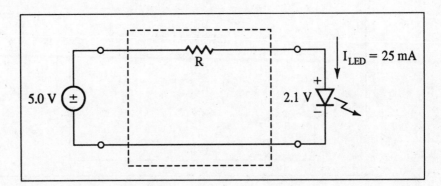

Checking the tables of available resistors, we find that 110 Ω is the nearest standard value 5% tolerance, carbon resistor, which happens also to be the least expensive. Because a resistor with a value higher than 116 Ω would limit the current to less than 25 mA, we chose the closest standard value lower than 116 Ω. At this point, our design is as shown in Fig. 2.3.

Step 4: Predict the Performance.

We analyze the circuit to predict the current through the LED, as follows:

$$I_{LED} = \frac{(5.0 - 2.1)}{110} A \qquad (2.2)$$

giving

$$I_{LED} = 26.36\, mA.$$

We also need to predict the power dissipated in the resistor, as follows:

$$P = 2.9\ V \bullet 26.36\ mA \qquad (2.3)$$
$$= 76.44\ mW.$$

In the process of creating the design, we chose a 5% tolerance resistor. We should note that our design neglects the possible effects on the performance resulting from that tolerance. We discuss dealing with component tolerances in Chapter 3. We also selected a 1/4-W carbon resistor, which will easily dissipate the predicted 76.44 mW. In fact, we could have chosen an 1/8-W resistor; however, it happens that the 1/8-W resistor is more expensive. Usually we choose the least expensive component unless there is some overriding reason to do otherwise; for example, we might select a 1/8-W resistor, although it is more expensive, for a space satellite application because it's smaller and lighter than a 1/4-W resistor.

While analyzing our design to predict its performance, we also verify that the design objectives have been met. In our example, our choice of circuit configuration and resistor value has resulted in a design that satisfies the performance requirements, addresses the design criteria, and doesn't violate any design constraints.

Finally, we need to document and present the final design for evaluation and eventual implementation. We discuss this step in greater detail in Section 2.4.

Note that if some or all of the objectives had not been met, we would need to return to an earlier step in the process and either change our circuit configuration or resistor value (for example, we might choose a more expensive 1% tolerance metal film resistor and specify a standard value of 115 Ω, which is very close to our desired 116 Ω). In

Figure 2.3. Proposed Design for Example 2.1.

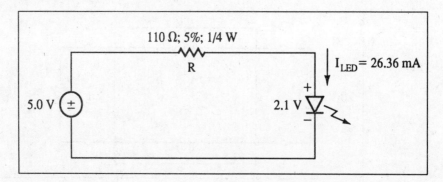

some cases, it might even be necessary to reconsider the design objectives and perhaps relax some of the requirements or constraints in order to achieve a viable design.

It is worth noting that as we developed our final design, we applied certain intellectual processes to the problem. First, we took time to interpret the problem as we set the design objectives. Then, as we selected a circuit configuration and resistor value, we worked at synthesizing a possible solution to the situation. Next, we analyzed our design to predict how it would behave. Finally, we evaluated the predicted behavior against our established objectives. Generally, all design projects will call on all of these intellectual processes, and we often have to juggle between them as we converge on an acceptable solution.

2.3 THE ENGINEER'S TOOLS

The analysis needed to solve the design problem in Example 2.1 was fairly straightforward; we selected that example so we could focus on the design process without becoming too immersed in detailed analytic procedures. However, because you will be applying this process to more challenging problems in which the analysis will be more intricate, let's review the tools that are available to you to facilitate the analytic process.

An electrical engineer is trained to understand the physics of fields, devices, and circuits and to analyze them mathematically. Usually the analysis uses a mathematically formulated model of a device or circuit to predict the performance of the real device or circuit. Often, the designer must choose between a simple model that is easily analyzed and a complex model that more accurately portrays reality but which may require numerical analysis by a computer.

Using models to predict performance is convenient and effective as part of the design process. However, because a model is never more than a useful approximation of reality, a prototype is usually built and tested. The results of these tests then can be coordinated with refinements to the model. (For obvious reasons, we have omitted the prototype phase of the process from the exercises in this compendium.)

Computers are particularly useful when the design process is iterative. In this case, a ready-for-prototyping design would be arrived at through a cycle of analyzing the performance of a candidate design, modifying as needed the assumptions, parameters, and structure, and then repeating the performance analysis. This repetitive analysis procedure requires solving many equations, a process that is best handled by a computer. Computer software exists that can greatly facilitate this activity by simulating (modeling) the performance of electrical devices and circuits. Examples of this software are PSPICE®, MicroCAP®, and MicroLogic™. Enhancements to these simulators can help the engineer in the areas of optimization, element sensitivity and variability, etc.

Other useful software includes products such as MathCAD® and TK!Solver™ which can aid general equation solving. Derive® is a similar product that manipulates symbolic representations of variables. Spreadsheet programs such as Lotus® 123®, Quattro® Pro, and Microsoft® Excel perform programmed computations and provide impressive graphics. Their strengths lie in their interactive nature and their ability to deal with multiple options within one spreadsheet. And sometimes, of course, it is appropriate for the engineer to write a simple, custom program in, say, BASIC or C.

2.4 PRESENTING A DESIGN FOR EVALUATION

Let's now turn our attention to the important topic of documenting designs and presenting them for evaluation. The evaluator of your design, typically your course instructor (or, in industry, your supervisor) should ensure your design meets certain requirements.

Specifically, the evaluator should examine your design's configuration and parameter values and its capability to meet the performance requirements and satisfy the design criteria without violating any design constraints. The evaluator should be concerned about the design's reliability and safety and whether it can actually be manufactured. Note that these requirements differ from the answers to exercises typically found in engineering textbooks. For your design to be evaluated fairly, however, you must present it so that the evaluator can easily understand it.

From an academic point of view, we address this topic from two angles: We deal first with the designer's responsibility and second with the evaluator's. In practice, however, these responsibilities are shared by the designer, who must present the design effectively and with appropriate technical support, and by the evaluator, who must judge the design's merits and sanction its implementation.

The Designer's Responsibility

Your final design is the principal tangible outcome of your engineering design activity and is evidence of your skill as a designer. When you prepare the final design of an electrical device or circuit for evaluation, you should assemble the documentation carefully and present it in the following format:

Title Information

Clause describing design activity
Design exercise reference number
Designer's name
Date

Design Objective

Design Objective Statement. Begin with a statement of the design objective expressed succinctly in engineering terms. Don't address the motivational or applications background of the exercise. Leave discussion of specific issues to the following three subsections. Usually, you will want to express the objective in a single sentence with few or no numbers.

Performance Requirements. Key to developing a satisfactory design is a careful and complete enumeration of the performance requirements. These requirements must be specific and relate to the required electrical (and occasionally mechanical) performance characteristics in terms of voltage, current, impedance, power, time, frequency, etc., at the input or output of your device or circuit. Because these requirements usually are a set of conditions that express the limits for the satisfactory performance of your design, often you can list the actual requirements in terms of simple mathematical inequalities, as illustrated in the following:

bandwidth, BW	$BW > 1\ MHz$
voltage gain, A_V	$17\ dB < A_V < 23\ dB$
or, alternatively,	
voltage gain, A_V	$A_V = 20 \pm 3\ dB.$

In many cases, the performance requirements are not stated explicitly in the exercise, but instead reference is made to named input or output standards. In these cases, you would need to ferret out the performance requirement details from published standards or data sheets and then list the relevant requirements in this section of the Design Objective.

Design Constraints. These constraints reflect those limitations on the design that are fixed prior to the onset of the design process; for example, use of specific parts or of a particular power supply voltage or limit on the allowable cost of the product. As designer, you generally will have no flexibility to modify these constraints. Note that the constraint to use approved parts only is generally implied in this compendium's exercises, so you can omit that one from your Design Objective Statement.

Design Criteria. These are the criteria against which competing designs are judged. Obviously, a design first must meet the performance requirements before it can be considered for competition. Criteria such as cost, reliability, efficiency, and parts count are often used; other criteria might include response time, bandwidth, or extent of an operating margin. When several important issues contribute to the quality of the design, optimization criteria are used that embody trade-offs expressed as sums, ratios, or products. Note also that because design is often conducted in an industrial or commercial environment, cost is always an unspoken criterion. Thus cost as a design criterion usually can be omitted from the Design Objective Statement unless its inclusion is specifically requested. In the exercise statements in this book, if no design criteria are mentioned, you may omit this category of the Design Objective.

Design Composition

Engineering Drawings, Materials, and Parts. If the design concerns a device, you would prepare an engineering working sketch that shows dimensions, material specifications, and assembly method. On the other hand, if the design concerns a circuit, you would include a standard symbolic circuit schematic that shows component specifications. If material or component specifications cannot be easily presented on a drawing, you can include a reference to a separate materials or components list. Usually, you want this segment of the presentation to contain detail sufficient for someone to be able to build the device or circuit. Note also that this segment of the presentation will be combined with the permanent design documentation of a larger system that has the device or circuit as one of its parts.

Predicted Performance, Satisfaction of Constraints, and Evaluation of Criteria. In this section, you describe the predicted performance of the device or circuit. The predicted performance shows the specific performance you expect from your design and should be presented in a way that enables direct comparison to show that this performance meets or exceeds the stated performance requirements. Depending on the particular design and its objectives, the performance might be described by a single number, a table, or a graph. In this section, you also should verify the design does not violate the stated constraints and confirm it meets the stated criteria.

Proof Of Performance

In this part, you blend the following related issues:
- The assumptions made during the course of the design
- The models that represent the behavior of the device or circuit
- An analysis or simulation based on these assumptions and models that predicts the performance of the device or circuit

In some design exercises, these issues are clearly separable and can be presented sequentially. In others, you might find it necessary to combine assumptions, models, and analysis as the argument in the Proof is developed. In either case, you should compile the Proof logically and include clear references to the meanings of notational terms. The Proof may take the form of a mathematical analysis or a computer simulation. Note that a Proof of Performance is proof that the device or circuit will perform as you predict; it is not a demonstration of how you arrived at the values of the components.

Design Synthesis Record (included at the discretion of the evaluator)

The chronological record of design synthesis can be appended if requested by the design's evaluator. The record usually shows the engineering method and the calculations you used and discusses decisions you made during the course of the design development. Customarily, design engineers keep the synthesis records of their designs in an Engineer's Notebook. While the records in these notebooks may seem superfluous once a design has been finalized and presented, they often become the legal record of invention when awarding patents or the legal evidence in the event of lawsuit concerning the design. Omitting this record from the presentation reduces the amount of material to be submitted and focuses the evaluation on the outcome of the design rather than on the process followed to develop it.

Having outlined the format of the design presentation, let's review it in the following two example exercises. The final design corresponding to Example 2.1, stated earlier, is presented in Fig. 2.5. Example 2.2, given below, is depicted in Fig. 2.6.

Example 2.2. Resistor Network.

Color television cameras capture the red, green, and blue shades of an image as separate signals. However, to retain compatibility with black-and-white television receivers, the red, green, and blue voltage signals (V_R, V_G, and V_B, respectively) are combined into a single black-and-white luminance (V_L) signal according to the function

$V_L = 0.30 V_R + 0.59 V_G + 0.11 V_B$.

The luminance signal, together with two chrominance signals, which are derived from similar functions of the red, green, and blue signals, are combined into a composite television broadcast signal. While black-and-white television receivers display only the luminance signal, color receivers reconstruct the red, green, and blue shades from the luminance signal and two chrominance signals.

Using 5% preferred resistor values only, complete the design of the following three-resistor network to generate the luminance signal from the red, green, and blue voltage signals according to the function given above. The circuit must implement the coefficients in the function as closely as possible while presenting an equivalent output resistance of 3000–4000 ohms. There is no load on the output of the circuit. See Fig. 2.4.

While Examples 2.1 and 2.2 differ considerably, the formats of their design presentations are similar. We have already discussed the Example 2.1 exercise in some detail and recognize that it might relate to an elementary circuits class. The goal of the exercise, the circuit performance requirements, and the particular design constraints are clearly set out in the Design Objective Statement. In the Design Composition, the circuit diagram is drawn in a standard symbolic form and the components are unambiguously specified. The predicted performance of the circuit is presented in accordance with the performance requirement, which is satisfied. The Proof of Performance is based upon elementary circuit theory.

Example 2.2 might relate to a beginning circuits course, where Kirchhoff's laws and resistive circuit analysis techniques are covered. The design is presented in the same form as in Example 2.1 except that it is slightly more elaborate. The Design

Figure 2.4. Design of a Three-Resistor Network.

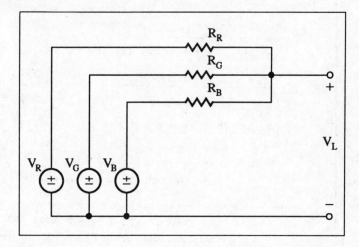

Composition presents the circuit with specified 5% resistor values and shows the predicted functional relationship between the television luminance output and the red, green, and blue input signals. The Composition also confirms that the predicted output resistance falls within the range specified in the Design Constraints. The first part of the Proof of Performance is a straightforward application of Kirchhoff's current law to derive an expression of the output voltage. Numerical values are inserted into this function based on the specified resistors, verifying the results presented under the predicted performance. The second part of the Proof addresses the design criterion of having the coefficients of the function match the desired coefficient values as closely as possible. An extract from a computer spreadsheet analysis is shown to illustrate the closeness of competing designs in matching the coefficients. An objective function (see Chapter 6, Optimization) was set up to judge each design.

The computer spreadsheet analysis is a useful, although not essential, tool in iterative designs of this kind. The same procedures could have been performed quite easily (but with more tedium) on an electronic calculator.

The Evaluator's Responsibility

An evaluator reviewing design presentations uses engineering expertise and experience to answer the following questions:

1. Does the designer have a thorough understanding of the purpose and goals of the design? Have all of the relevant requirements, constraints, and criteria been identified?
2. Is the general configuration of the device or circuit plausible for meeting the design objectives? Do the device dimensions, material specifications, or circuit component values appear to be within the expected range?
3. Does the predicted performance of the device or circuit meet or exceed the performance requirements? Can this level of performance be expected from the proposed device or circuit, given the specified dimensions, materials, and component values? Has the designer taken into consideration the constraints? Have the merits of the design been evaluated appropriately in terms of the design criteria?
4. What assumptions were made in the course of the design and are these assumptions valid? What body of theory has the designer employed and is this theory relevant to the exercise?
5. Are there any flaws in the analysis done in the Proof of Performance?

Figure 2.5. The Design Presentation for Example 2.1.

1. DESIGN OBJECTIVE

   DESIGN A CIRCUIT TO DELIVER $\geq 25\,mA$ TO A LED.

   PERFORMANCE REQUIREMENTS

   * LED CURRENT   — $\geq 25\,mA$

   DESIGN CONSTRAINTS

   * SUPPLY VOLTAGE   —   5.0 V
   * VOLTAGE DROP ACROSS LED —   2.1 V
         (FOR LED CURRENT IN RANGE 25-30 mA)

   DESIGN CRITERIA

   * LED CURRENT TO BE AS CLOSE TO 25 mA AS POSSIBLE, AND NOT TO EXCEED 30 mA.

2. DESIGN COMPOSITION

   CIRCUIT SCHEMATIC:

   110 Ω 5%
   ¼ W
   5.0 V

   PREDICTED PERFORMANCE:

   * LED CURRENT: 26.36 mA

EXAMPLE 2.1
LED INDICATOR CIRCUIT

NIGEL T. MIDDLETON
10-24-89
PAGE 1 OF 2

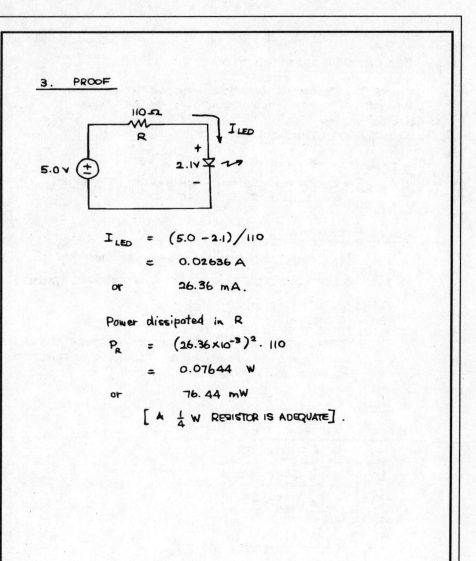

3. PROOF

$$I_{LED} = (5.0 - 2.1)/110$$
$$= 0.02636 \text{ A}$$
or
$$26.36 \text{ mA}.$$

Power dissipated in R
$$P_R = (26.36 \times 10^{-3})^2 \cdot 110$$
$$= 0.07644 \text{ W}$$
or
$$76.44 \text{ mW}$$
$$\left[A \ \tfrac{1}{4} \text{ W RESISTOR IS ADEQUATE} \right].$$

MIDDLETON / EX 2.1   PAGE 2 OF 2

Figure 2.5. The Design Presentation for Example 2.1. (Continued)

Figure 2.6. The Design Presentation for Example 2.2.

1. DESIGN OBJECTIVE

COMPLETE THE DESIGN OF RESISTOR NETWORK TO GENERATE
LUMINANCE FROM RGB VOLTAGE SIGNALS, BY SPECIFYING
RESISTOR VALUES.

PERFORMANCE REQUIREMENT

$$V_L = 0.30 \cdot V_R \pm 0.59 \cdot V_G + 0.11 \, V_B$$

SUBSCRIPTS. IDENTIFY LUMINANCE, RED, GREEN
AND BLUE VOLTAGE SIGNALS.

DESIGN CONSTRAINTS

1. EQUIV. OUTPUT RESISTANCE $3000 \leq R_{OUT} \leq 4000 \, \Omega$

2. RESISTORS SELECTED FROM   PREFERRED VALUE.

DESIGN CRITERIA

COEFFICIENTS IMPLEMENTED BY DESIGN MUST MATCH
COEFFICIENTS IN PERFORMANCE REQUIREMENT FUNCTION
AS CLOSELY AS POSSIBLE.

2. DESIGN COMPOSITION

R_R: 11KΩ, 5%
R_G: 5.6KΩ, 5%
R_B: 30KΩ, 5%

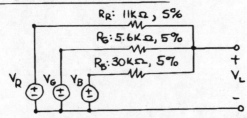

PREDICTED PERFORMANCE:

$$V_L = 0.3002 \, V_R + 0.5897 \, V_G + 0.1101 \, V_B$$
$$R_{OUT} = 3302 \, \Omega$$

EXAMPLE 2.2.

LUMINANCE / RGB NETWORK

NIGEL T. MIDDLETON
7-23-91
PAGE 1 OF 3

3. PROOF

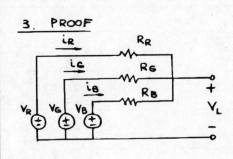

KIRCHOFF'S CURRENT LAW:

$$i_R + i_G + i_B = 0$$

WHERE

$$i_R = \frac{V_R - V_L}{R_R}$$

$$i_G = \frac{V_G - V_L}{R_G}$$

$$i_B = \frac{V_B - V_L}{R_B}$$

$$\therefore \quad \frac{V_R - V_L}{R_R} + \frac{V_G - V_L}{R_G} + \frac{V_B - V_L}{R_B} = 0$$

$$\text{OR} \quad V_L \left(\frac{1}{R_R} + \frac{1}{R_G} + \frac{1}{R_B} \right) = \frac{V_R}{R_R} + \frac{V_G}{R_G} + \frac{V_B}{R_B}$$

DEFINE EQUIVALENT OUTPUT RESISTANCE $R_{OUT} = R_R /\!/ R_G /\!/ R_B$

$$\text{OR} \quad \frac{1}{R_{OUT}} = \frac{1}{R_R} + \frac{1}{R_G} + \frac{1}{R_B}$$

SUBSTITUTING:

$$V_L = \frac{R_{OUT}}{R_R} \cdot V_R + \frac{R_{OUT}}{R_G} \cdot V_G + \frac{R_{OUT}}{R_B} \cdot V_B$$

DESIGN VALUES $R_R = 11000\,\Omega$; $R_G = 5600\,\Omega$; $R_B = 30000\,\Omega$

$$\therefore \qquad = 3302.4\,\Omega$$

AND $\frac{R_{OUT}}{R_R} = 0.3002$; $\frac{R_{OUT}}{R_G} = 0.5897$; $\frac{R_{OUT}}{R_B} = 0.1101$

$$\therefore \quad V_L = 0.3002\,V_R + 0.5897\,V_G + 0.1101\,V_B$$

FOLLOWING SPREADSHEET SUMMARIZES DESIGN ITERATIONS FOR R_{OUT} RANGING FROM $4000\,\Omega$ TO $3100\,\Omega$. THE "PRECISE" COLUMN SHOWS DESIRED $\quad$ S AND COEFFICIENTS (SHADED), AND REQUIRED RESISTOR VALUES. THE "5% PREF VAL COLUMN" SHOWS THE INSERTION OF CLOSEST 5% RESISTOR VALUES, WITH CONSEQUENT R_{OUT} AND COEFFICIENT VALUES. THE OBJECTIVE FUNCTION (MINIMUM IS BEST) RATES THE DESIGN WITH RESPECT TO COEFFICIENTS BY

$$\text{OBJ. FUNC} = (0.3 - C_R)^2 / 0.3^2 + (0.59 - C_G)^2 / 0.59^2 + (0.11 - C_B)^2 / 0.11^2$$

DESIGN # 4 IS BEST.

MIDDLETON / PAGE 2 OF 3

Figure 2.6. The Design Presentation for Example 2.2. (Continued)

Design iterations		Precise	5% Pref. value
Design 1	R out	4000	3972.0
	R (red)	13333.333	13000
	R (green)	6779.661	6800
	R (blue)	36363.636	36000
	Coeff. (red)	0.30	0.3055
	Coeff. (green)	0.59	0.5841
	Coeff. (blue)	0.11	0.1103
	Objective fn.	0	4.4967E-04
Design 2	R out	3700	3637.3
	R (red)	12333.333	12000
	R (green)	6271.186	6200
	R (blue)	33636.364	33000
	Coeff. (red)	0.30	0.3031
	Coeff. (green)	0.59	0.5867
	Coeff. (blue)	0.11	0.1102
	Objective fn.	0	1.4355E-04
Design 3	R out	3500	3422.2
	R (red)	11666.667	12000
	R (green)	5932.203	5600
	R (blue)	31818.182	33000
	Coeff. (red)	0.30	0.2852
	Coeff. (green)	0.59	0.6111
	Coeff. (blue)	0.11	0.1037
	Objective fn.	0	6.9953E-03
Design 4	R out	3300	3302.4
	R (red)	11000.000	11000
	R (green)	5593.220	5600
	R (blue)	30000.000	30000
	Coeff. (red)	0.30	0.3002
	Coeff. (green)	0.59	0.5897
	Coeff. (blue)	0.11	0.1101
	Objective fn.	0	1.2686E-06
Design 5	R out	3100	3002.0
	R (red)	10333.333	10000
	R (green)	5254.237	5100
	R (blue)	28181.818	27000
	Coeff. (red)	0.30	0.3002
	Coeff. (green)	0.59	0.5886
	Coeff. (blue)	0.11	0.1112
	Objective fn.	0	1.2171E-04

Figure 2.6. The Design Presentation for Example 2.2. (Continued)

Above all, the evaluator should search for reasons why the design, and consequently the device or circuit, might fail. This is why it is important for the designer to present the design clearly, thoroughly, and accurately. For instance, when evaluating the designs in Examples 2.1 and 2.2, the evaluator should be convinced that the designer understands the objectives and that the device and circuit configurations are viable for the design objectives. This conviction should be supported by the performance predictions, which should be clearly laid out and supported in the proofs. Finally, the designs should be presented with the conviction that they will work so that the evaluator will more likely have confidence in the designer's work.

The methods of presentation described in this chapter are consistent with those often applied in the engineering industry. At Texas Instruments[1], for example, a formal design review procedure exists to comprehensively address all aspects of a new product design before releasing it for fabrication. A key ingredient in this procedure is the written documentation pertaining to new designs. The first level in TI's review procedure is the Internal Critical Design Review. In this review, documentation presented by the design engineer includes design specifications and requirements, interface requirements, schematics and block diagrams, theory of operation, worst-case performance analysis, simulation results, and a bill of materials. The system engineer then will use this documentation to verify the adequacy of the design with respect to such issues as the nominal and worst-case electrical function, the performance against requirements, power, reliability, safety, and parts availability, to mention only a few of the more pertinent issues. After the design passes this internal review, it is then reviewed again with the client in attendance, where once again a similar detailed set of design documentation is submitted.

---

[1]According to Mr. Phil Congdon, Manager, Systems Technology Laboratories, Texas Instruments, Inc.

3

Designing for Mass Production

"Close only counts in horseshoes."
Anonymous

3.1 DESIGNING WITH REAL COMPONENTS

In the physical world, often it turns out that something is not exactly the dimension or value we say it is. For example, when we measure and cut a board to a length of 4 ft, the result might be at best only close to the desired goal. A competent carpenter can cut the board to within 1/8 in. In this case, the board would be cut to a nominal value, or length, of 4 ft with a tolerance of 1/8 in. The *nominal value* is the target value—the length, in this case—while the *tolerance* is the amount of error from the nominal value that is considered satisfactory and still produces the desired result. In the situation of a cut board, a 1/8-in. tolerance usually is satisfactory for framing a house, but too much for a piece of fine furniture.

A similar situation exists for all manufactured components. For example, carbon resistors are typically manufactured with a tolerance of 5 or 10%; for example, a resistor with the markings BROWN-BLACK-RED-GOLD (1000-Ω nominal value, 5% tolerance—see Part 2) will have the actual value of 950–1050 Ω (1000 Ω, ±5%).

In addition to the variable of manufacturing tolerance, component values also vary with temperature, humidity, age, mechanical stress, and other factors. For example, a metal film resistor with a 1% tolerance at 25°C can have a temperature coefficient as large as ±100 ppm/°C (parts per million per °C). This means that if the temperature changes by 70°C, the resistance may change by ±0.7% from its original value, because

$$70 \cdot \frac{100}{10^6} = 0.007, \text{or } 0.7\%.$$

Component variations due to these factors must be combined with the manufacturing tolerance to determine the range of values a component can actually take on. For a 1000-Ω (nominal value), 1% metal-film resistor operating over a temperature range of −45 to +95°C (25°C, ±70°C), its resistance value can be

$983.07 < R < 1017.07 \ \Omega$.

The minimum value is given by

$$R_{min} = R_{nom} \ (1 - .01) \ (1 - .007) \ \Omega,$$

and the maximum value is given by

$$R_{max} = R_{nom} \ (1 + .01) \ (1 + .007) \ \Omega.$$

Note that the variations due to the tolerance and the temperature coefficient are multiplied, not added. Component variations due to other factors are handled in a similar manner.

Engineers specify components and materials that are procured or manufactured to be assembled into larger devices or systems. The actual value of these components (dimension, resistance, capacitance, etc.) may vary from their nominal value, yet the finished device must operate within specified performance requirements.

If the device you are designing is to be manufactured in large quantities, a random selection of components placed in the circuit will produce a random variation in the performance of the circuit. For example, if two resistors, each with 5% tolerance, are connected in series with a voltage source, the voltage at the node between the two resistors can differ from the nominal value by a significant amount. (The actual amount depends on the ratio of the two resistance values.) If this voltage divider is used to bias a transistor amplifier, for example, this variation could have a significant and maybe catastrophic effect on the operation of the amplifier. Tolerances as well as temperature-induced variations in the transistor parameters will augment these effects on the amplifier operation. You would need to ensure that this amplifier circuit will operate despite these variations. If the circuit were to be manufactured in large numbers, all possible combinations of parameter values might occur and must be expected. Repairing, potentiometer adjustments (trimming), and hand-selecting to match parts are expensive processes on a production line and are poor substitutes for a design that accommodates these component variations.

Designing with nominal values alone does not guarantee that all finished devices or circuits will meet performance requirements. In the next sections, we discuss how to deal with component variations in a design.

3.2 DESIGNING WITH NOMINAL COMPONENTS

Example 3.1 is used in this section and the next to illustrate designing with real components. In this section we do a nominal design and in the next section we illustrate three methods of including the effects of component variations.

Example 3.1. A Voltage Divider.

Design the voltage divider shown in Fig. 3.1 so that the output voltage will be 2.0 ±0.4 V. The voltage source available may vary by ±5% and the resistors available are carbon, standard values with 5% tolerance. Because of power consumption constraints, the current from the power supply must be kept below 1 mA. Assume there is no load on the circuit. This circuit will be used in a consumer product with a temperature range of 0-70°C.

Performance Requirement: $V_o = 2.0 \pm 0.4 \ V$

over a temperature range 0-70°C

Design Constraints: Must use 5% carbon resistors

Supply voltage $= 5.0 \ V \pm 5\%$

Supply current $< 1.0 \ mA$

Circuit configuration given

Figure 3.1. Voltage Divider Network for Example 3.1.

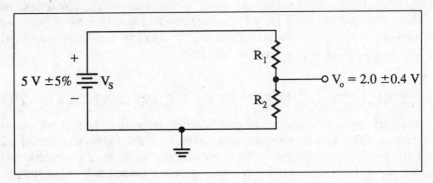

To design this circuit, you could simply try choosing some values of resistors and then analyzing the circuit to see if performance requirements are met, obviously a very time-consuming approach. Another, more direct approach would be to select resistor values based on the performance requirements and then verify by analysis that the proposed design meets those requirements.

The equation for the voltage at the output is

$$V_o = V_s \frac{R_2}{R_1 + R_2}. \tag{3.1}$$

We have two unknowns but only one equation. Given no other constraints, you would need to use your judgment and experience to develop another relationship that allows the solution for two unknowns. For this example, the source current must be kept below 1 mA, which means that the sum of the two resistors must be greater than 5.25 kΩ. [5.25/($R_1 + R_2$)<1 mA.] For instance, you could decide that the sum of the two resistors should be approximately 10 kΩ. This series resistance would limit the source current to 0.5 mA when all components are nominal and would accommodate considerable variation due to tolerance and temperature factors, while still staying within the 1 mA constraint. Obviously, many other assumptions could be made here and many other solutions are possible; this is where a designer must use judgment. As you will see when we get to a final design, this assumption is very conservative.

Starting with the initial trial that $R_1 + R_2 = 10$ kΩ, we find that a nominal solution would be to set $R_1 = 6$ kΩ and $R_2 = 4$ kΩ. However, these are not standard values available with 5% tolerance. If we choose R_2 to be 3.9 kΩ, the nearest standard 5% value (see Part 2), then we can solve Eq. 3.1 for R_1 as follows:

$$R_1 = R_2 \left(\frac{V_s}{V_o} - 1 \right) = 5.85 \text{ k}\Omega. \tag{3.2}$$

Once again, this is not a standard 5% value. The nearest standard value is 5.6 kΩ. Using 5.6 kΩ, the nominal output voltage would be

$$V_o = \frac{5.0 \text{ V} \bullet 3.9 \text{ k}\Omega}{5.6 \text{ k}\Omega + 3.9 \text{ k}\Omega} = 2.053 \text{ V}, \tag{3.3}$$

and the nominal source current would be

$$I_s = \frac{5.0 \text{ V}}{5.6 \text{ k}\Omega + 3.9 \text{ k}\Omega} = 0.526 \text{ mA}, \tag{3.4}$$

both of which are well within specification. Therefore our proposed design is $R_1 = 5.6$ kΩ, $R_2 = 3.9$ kΩ, carbon-composition, 5% resistors. Power dissipation calculations show that 1/4-W power dissipation capacity is adequate (5.0 V $\bullet$ 0.526 mA = 2.63 mW total).

Our task is not finished, however. So far, we have used only the nominal values for our calculations. What will happen if we want to manufacture a large number of these circuits? Will they all meet the performance requirements? Next, we discuss methods to verify that the manufacturing process can consistently turn out circuits that will meet the performance requirements.

3.3 INCLUDING THE EFFECTS OF COMPONENT VARIATIONS

There are three methods available for evaluating the design: The first two, the worst-case method and the variation-of-parameters method, are deterministic; the third, the Monte-Carlo method, is probabilistic. The worst-case and variation-of-parameters methods are very conservative and describe the absolute performance limits of the circuit when the values of components at their extreme variation occur. The Monte-Carlo method, on the other hand, will predict the expected distribution of performance values when a large number of circuits are made with random selections of components.

Worst-case Analysis

A worst-case analysis assumes that the worst possible combinations of component values can occur. The designer seeks to guarantee that even if these combinations do occur, the circuit or device will still meet performance requirements.

Equation 3.1 expresses the output voltage (the performance value) of the circuit in terms of the circuit components. To determine the worst-case limits of the output voltage, we analyze the circuit with the combinations of component values that will give the worst-case output voltages, both minimum and maximum.

Worst-case Resistance Values

Before we discuss the process, let's take a necessary side trip to determine the actual worst-case limits on the values of the resistors.

In Example 3.1, we are limited to using 5% tolerance, carbon resistors for the voltage divider. If we measure the actual resistance of a resistor selected at random, it would fall within the limits shown in Table 3.1.

The operating temperature compounds the manufacturing tolerance. The temperature characteristics for carbon resistors is given in Part 2. The appropriate parts of the temperature characteristic table are transferred to Table 3.2. The lowest temperature of concern is $0°C$, the highest $70°C$. We have to interpolate between 65 and $85°C$ to get the characteristic for $70°C$.

The resistance values of interest are $1000-10,000\ \Omega$, so we will again use linear interpolation to approximate the intermediate values. The results of the interpolation are shown in Table 3.3.

Table 3.1. Resistor Values. Minimum and maximum values of the 3.9 kΩ and 5.6 kΩ resistors in Example 3.1 due to 5% manufacturing tolerance (at 25°C).

Min (-5%)	Nominal	Max (+5%)
3705	3900	4095
5320	5600	5880

Table 3.2. Resistor Temperature Variations. Portions of the temperature characteristics data for carbon-composition resistors, along with the interpolation for 70°C. This table gives the maximum temporary resistance change (in percent) from the initial resistance value at +25°C.

Temp 0°C	+25° C Nominal	+65.°C	+85°C	+70°C
-0.5– + 1.4	1,000	-1.2– + 1.8	-1.3– + 3.3	-1.225– + 2.175
-0.6– + 1.7	10,000	-1.4– + 2.2	-1.5– + 3.9	-1.425– + 2.625

Note that the temperature characteristic has both positive and negative values. The range indicates that the resistance value of any particular resistor may rise or fall with temperature, independent of other resistors. Also note that the maximum resistance change or drift is greater in both directions at 70°C than at 0°C. Thus only the higher temperature need be considered in this example.

Table 3.3. Worst-case Limits. Interpolation of Table 3.2 for 3.9 kΩ and 5.6 kΩ and the worst-case limits of the resistors, including the 5% manufacturing tolerance. (R_1min = 5600 • 0.95 • 0.98663 Ω)

R	Nominal	% Change at 70°C	R_{min}	R_{max}
R1	5600 Ω	-1.337– + 2.427	5249 Ω	6023 Ω
R2	3900 Ω	-1.303– + 2.351	3657 Ω	4191 Ω

Now we return to worst-case performance calculations.

Worst-case Maximum

Clearly, from looking at Eq. 3.1, we can see that to get the maximum output voltage, we should take V_s at its maximum and R_1 at its minimum value. However, it's not immediately clear which limit to use for R_2, since it appears in both the numerator and the denominator in Eq. 3.1. When in doubt, we can take the derivative with respect to the parameter in question, R_2. The sign of the derivative will tell us if the output voltage will increase or decrease with increasing R_2.

To calculate the maximum output voltage, we use the maximum value of R_2 if the sign of the derivative is positive and the minimum value of R_2 if the sign is negative. For the minimum output voltage calculation, we use the reverse selection. Nominal values for the parameters are usually sufficient when determining the derivative sign.

Taking the partial derivative with respect to R_2,

$$\frac{\partial V_o}{\partial R_2} = \frac{V_s (R_1 + R_2) - (V_s R_2)}{(R_1 + R_2)^2}$$

$$= \frac{V_s R_1}{(R_1 + R_2)^2} \tag{3.5}$$

Since this derivative is positive, the maximum value of R_2 will give a maximum output voltage.

For this example, worst-case analysis reveals

$$V_{out\,max} = \frac{V_{s\,max} \bullet R_{2\,max}}{R_{1\,min} + R_{2\,max}} = 2.331 \text{ V } (=2.0\text{ V} + 16.6\%). \tag{3.6}$$

Similarly,

$$I_{s\,max} = \frac{V_{s\,max}}{R_{1\,min} + R_{2\,min}} = 0.589 \text{ mA}. \tag{3.7}$$

Worst-case Minimum

To obtain the minimum value of output voltage, use the opposite parameter values as were used for the maximum calculation, as follows:

$$V_{out\,min} = \frac{V_{s\,min} \bullet R_{2\,min}}{R_{1\,max} + R_{2\,min}} = 1.794 \text{ V} \quad (=2.0\text{ V} - 10.3\%). \tag{3.8}$$

Because all the performance predictions are within the allowable specifications, this design meets specifications under the worst-case conditions. Note, however, that because a minimum value of I_s was not specified, this calculation wasn't necessary. Also, when the maximum current was calculated, a different combination of minima and maxima was used than for the output voltage.

Analysis by Variation of Parameters

The variation-of-parameters method of analysis is similar to the worst-case method. It proceeds from the nominal design procedure shown in Section 3.2. Partial derivatives of the output voltage with respect to each component are used to determine the relative effect of variations in each component value on the output voltage. The derivatives are calculated at nominal values and are known as *sensitivity coefficients*. The variation of the output voltage caused by variation in the component value in question is the product of the sensitivity coefficient and the component variation. The total expected variation in the output voltage due to variations in all component values is the sum of the variation caused by each component. The variation in the output voltage of the example is expressed as

$$\Delta V_o = \frac{\partial V_o}{\partial V_s} \bullet \Delta V_s + \frac{\partial V_o}{\partial R_1} \bullet \Delta R_1 + \frac{\partial V_o}{\partial R_2} \bullet \Delta R_2, \tag{3.9}$$

where the ΔV_s, ΔR_1, and ΔR_2 are the variations from the nominal values of the corresponding components.

Note that each component normally has both a positive and a negative variation. Also, some of the sensitivity coefficients may be positive, while others are negative. To determine the maximum value of the performance parameter, use positive component variation with positive sensitivity coefficient and the negative component variation with negative sensitivity coefficient. To determine the minimum value of the performance parameter, use the opposite pairing.

For the purposes of this discussion, we will consider only the maximum variations from the nominal and simply call this the variation. We are interested in both positive and negative component variations. For example, for the two resistors in Example 3.1, the component variations are given in Table 3.4.

The derivatives are as follows:

$$\frac{\partial V_o}{\partial V_s} = \frac{R_2}{R_1 + R_2} \tag{3.10}$$

Table 3.4. Component Variations. Data for the resistors and the voltage source over their entire operating range.

	Min	Nom	Max	Neg Variation $(\Delta -)$	Pos Variation $(\Delta +)$
R_1	5249	5600	6023	-351 Ω	+423 Ω
R_2	3657	3900	4191	-243 Ω	+291 Ω
Vs	4.75	5.00	5.25	-0.25 V	+0.25 V

$$\frac{\partial V_o}{\partial R_1} = \frac{-V_s R_2}{(R_1 + R_2)^2} \tag{3.11}$$

$$\frac{\partial V_o}{\partial R_2} = \frac{V_s(R_1 + R_2) - V_2 R_2}{(R_1 + R_2)^2} = \frac{V_s R_1}{(R_1 + R_2)^2}. \tag{3.12}$$

When these sensitivity coefficients are evaluated for nominal component values,

$$\frac{\partial V_o}{\partial V_s} = 0.4105 \tag{3.13}$$

$$\frac{\partial V_o}{\partial R_1} = -0.21607 \bullet 10^{-3} \text{, and} \tag{3.14}$$

$$\frac{\partial V_o}{\partial R_2} = 0.31025 \bullet 10^{-3}. \tag{3.15}$$

The partial derivatives with respect to V_s and R_2 are positive, while the partial derivative with respect to R_1 is negative.

By the variation-of-parameters method, the positive variation of the output voltage is

$$\Delta V_{o_{pos}} = 0.4105 \bullet 0.25 + (-0.21607 \bullet 10^{-3}) \bullet (-351) + 0.31025 \bullet 10^{-3} \bullet 291$$

$$= 0.1026 + 0.0758 + 0.0903$$

$$= 0.269 \text{ V} \tag{3.16}$$

and the negative variation is

$$\Delta V_{o\,neg} = 0.4105 \bullet (-0.25) + (-0.21607 \bullet 10^{-3}) \bullet 423 + 0.31025 \bullet 10^{-3} \bullet (-243)$$

$$= -0.1026 - 0.0914 - 0.0754$$

$$= -0.269 \text{ V}. \tag{3.17}$$

Because the nominal output voltage is 2.053 V,

$$V_{o\,max} = 2.322 \text{V} \tag{3.18}$$

and

$$V_{o\,min} = 1.784 \text{V}. \tag{3.19}$$

These results are similar to but not quite the same as those of the worst-case analysis. The difference is that the worst-case method uses the actual values at the limits for the calculations, while the variation-of-parameters method relies on derivatives. The derivatives are calculated at nominal values, thus ignoring the fact that the derivatives will change over the full range of variation. The variation-of-parameters method, therefore, uses a straight-line approximation to a curve; Fig. 3.2 shows an

Figure 3.2. A Straight-line Approximation to a Curve for One Variable.

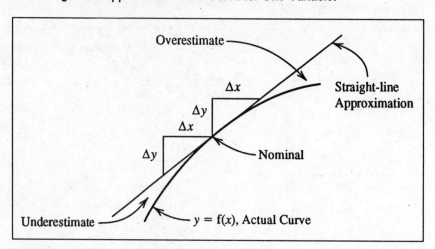

example of this approximation for one variable. In one direction, this approximation underestimates the variation; in the other, it overestimates the variation. The smaller the parameter variation, Δx, the better the approximation to Δy.

The variation-of-parameters method provides a significant result not available from the worst-case method. The sensitivity of the output to the variation of each parameter value is apparent from the calculations. While this example does not show great disparities in effect on the output voltage caused by each component, other examples would. You can use this information to improve a design; for example, if you use components with tighter tolerances or lower temperature characteristics, the output variation can be reduced. The "better" components are almost always more expensive; therefore a designer usually would use these components only in those applications in which there are large sensitivity coefficients.

Monte-Carlo Method

A Monte-Carlo analysis is a computer simulation of the production process. During production, each component is drawn at random from bins or boxes of the specified parts. Referring to Example 3.1, we can see that in addition to having a 5% manufacturing tolerance, the two resistors will experience a range of temperatures, expanding the range of values they may take on, as shown in Table 3.3. In a Monte-Carlo simulation, the program selects the value of each component at random within the limits of the component variation, simulating a random selection during manufacturing and subsequent operation at the temperature limits. System performance is then analyzed by the program using these random values. This process is repeated many times and the results plotted or otherwise presented to show the percentage of circuits whose performance meets the requirements under the full range of operating temperatures.

The Monte-Carlo method is illustrated for Example 3.1 in Figures 3.3, 3.4, and 3.5. Figure 3.3 shows a BASIC program used to calculate the output voltage of each circuit for the simulated production of 10 circuits. This program generates a random value for each component before each calculation, thus simulating a random selection of components during the manufacturing process. There are four steps in the process, as follows:

1. The random number generators are seeded (line 140), with each given a different starting value, one for each variable circuit component.

Figure 3.3. A BASIC Program for a Monte-Carlo Simulation of the Resistor Divider Circuit for Example 3.1. Also shown is the output listing of the first 10 trials.

```
10    REM MONTE CARLO ANALYSIS FOR A RESISTOR-DIVIDER CIRCUIT.
20    REM
30    REM NOTE THAT RND(A) PROVIDES A RANDOM NUMBER
40    REM BETWEEN 0 AND 1.
50    REM
60    REM PRINT HEADER
70    REM
80    LPRINT "TRIAL #          V SOURCE          R1          R2          V OUT"
90    LPRINT
100   REM
110   REM #140 SEEDS THREE RANDOM NUMBER GENERATORS (A, B, C) SO THE
120   REM RANDOM SELECTION OF EACH COMPONENT WILL BE INDEPENDENT.
130   REM
140            A=1:B=2:C=3
150   REM
160   REM THE FOLLOWING FOR-NEXT LOOP SIMULATES THE PRODUCTION
170   REM OF THE CIRCUIT WITH RANDOMLY SELECTED COMPONENTS.
180   REM
190            FOR N=1 TO 10
200   REM
210   REM SELECTING RANDOM VALUES FOR VS, R1, AND R2.
220   REM
230            VSN = 4.75+(.5*RND(A))
240            R1N = 5249+(RND(B)*774)
250            R2N = 3657+(RND(C)*534)
260   REM
270   REM DETERMINE OUPUT WITH THE RANDOMLY SELECTED COMPONENTS.
280   REM
290            VON = VSN*R2N/(R1N+R2N)
300   REM
310   REM NOW PRINT THE RESULT.
320   REM
330            LPRINT N,VSN,R1N,R2N,VON
340            NEXT N
```

TRIAL #	V SOURCE	R1	R2	V OUT
1	4.810675	5753.541	4120.972	2.007659
2	5.114881	5867.312	3696.355	1.9769
3	4.995157	5600.798	3714.271	1.991758
4	5.225255	5793.796	3941.016	2.115379
5	5.235581	5497.402	4167.572	2.257602
6	5.217258	5663.041	3958.401	2.146456
7	5.08561	5792.791	4052.574	2.093351
8	5.083439	5600.35	3835.433	2.066303
9	4.828427	5818.873	3946.898	1.951439
10	4.962985	5291.917	4067.255	2.156785

2. A random value is selected for each component (lines 230–250). The random number generator produces a number (uniformly distributed) between 0 and 1. The random component value is developed by multiplying the random number by the full range of component variation (max-min) to get a random variation within the interval. This variation then is added to the minimum component value.
3. The output performance value—the output voltage—is calculated (line 290).
4. The results of the calculation are printed (line 330).

The simulated manufacture of the circuit can be repeated any number of times in the FOR-NEXT loop (lines 190–340). Seeding the random number generators occurs only once, before the loop.

Figure 3.4 is a scatter diagram that shows the output voltages for the first 50 devices produced on our simulated production line. Figure 3.5 shows the distribution in the output voltage for a simulation of 10,000 circuits. Note that the output voltages tend to cluster around the value predicted for the nominal components, although some circuits appear near the limits. It can be shown that the distribution in performance characteristics will tend toward a Gaussian, or bell-shaped, distribution regardless of the distribution of the individual parts. This phenomenon is known in the field of statistics as the *Central Limit Theorem.*

We haven't discussed the distribution of component values within the variation limits. The program in Fig. 3.3 assumes a uniform distribution over the entire range. If another distribution is appropriate, you can modify the random number selection process in the simulation program.

3.4 CONCLUSION

Which analysis method should you use for these exercises? The answer will, of course, depend on the situation. Generally, a design meeting worst-case conditions is preferred; if a worst-case analysis shows that your design will perform as required, then that's all

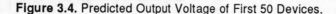

Figure 3.4. Predicted Output Voltage of First 50 Devices.

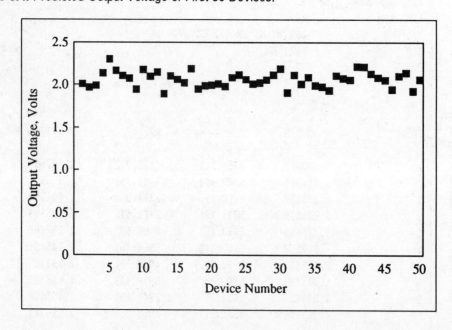

Figure 3.5. Output Voltage Distribution of 10,000 Completed Devices.

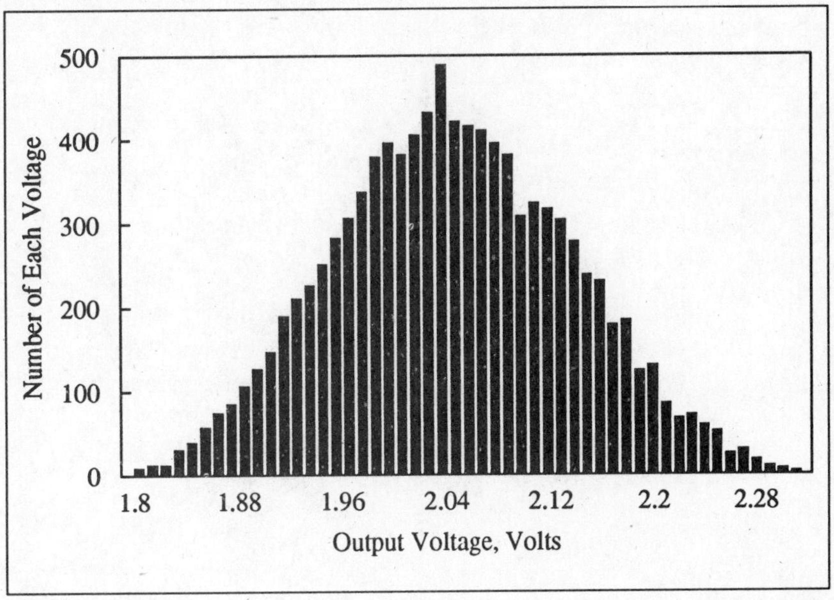

you need to do. In many cases, however, the worst-case limits may be slightly outside of the required specifications, so a Monte-Carlo analysis would be warranted. A Monte-Carlo analysis could show that the manufacturing yield might be high enough so that the total cost is less than the cost when using more expensive components to meet the worst-case conditions. In other cases, the circuit or system might be so complex that a set of equations could be too difficult to develop and a circuit simulator such as PSPICE®would be needed. Such simulators usually will provide sensitivity calculations that you can use to perform a variation-of-parameters method of analysis. In such a situation, you also could perform worst-case simulations, but doing so could become quite tedious.

We have used only manufacturing tolerance and temperature variation as examples of component variations. There are, in fact, many other causes of component variations, such as humidity, mechanical stress, and aging. The effects of these are significant; most designers consider them in the design process, although different designers might combine them in different ways. However, even though these variations will affect the magnitude of component variations, they won't affect the procedures we discussed in this chapter.

3.5 BIBLIOGRAPHY

1. Shooman, M. L. *Probabilistic Reliability: An Engineering Approach*. New York: McGraw-Hill, 1968.

2. Spence, Robert, and Randeep Singh Soin. *Tolerance Design of Electronic Circuits*. Reading, Mass: Addison-Wesley, 1988.

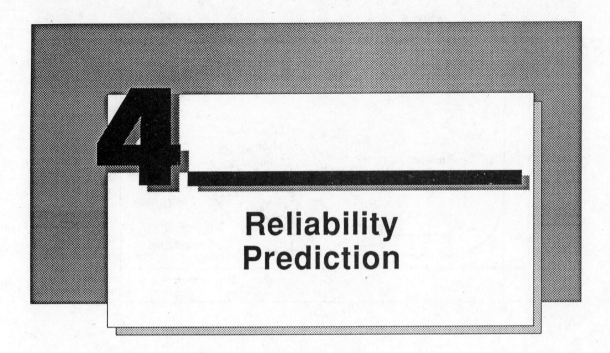

4

Reliability Prediction

"For loss of a nail, a shoe was lost.
For loss of a shoe, a horse was lost.
For loss of a horse, a rider was lost.
For loss of a rider, a battle was lost.
For loss of a battle, a kingdom was lost.
All for the loss of a nail."

Poor Richard's Almanac and Folklore

4.1 INTRODUCTION

An important performance characteristic of a system is its reliability. *Reliability* is the probability that a given circuit, device, or system will not fail over a specified time period. The initial purchase price of a system is only part of the cost of ownership; maintenance costs often play a major role in the overall, or life-cycle, cost. However, consistent, reliable service can be even more important than cost. For example, a breakdown to the military means "loss of mission," a situation that can have far-reaching consequences. For this reason, reliability of defense systems is very important. Even in consumer applications, loss of service can have more than monetary ramifications. For example, an electronic ignition module for a car might cost only about $100 to replace, but if the module were to fail in a remote location, the results could mean serious inconvenience or possibly even danger for the car's occupants.

4.2 BASIC CONCEPTS OF RELIABILITY

Reliability is measured in terms of failure. A *failure* is a condition in which a system or component loses its ability to fulfill its intended function. When a system ceases to operate within design specifications, it is considered to have failed. Similarly, an electrical or electronic component that doesn't operate within manufactured specifications also is considered to have failed; in practice, a failed electrical component usually

Figure 4.1. Bathtub Curve Showing Failure Rate Characteristics Over the Life of a Component.

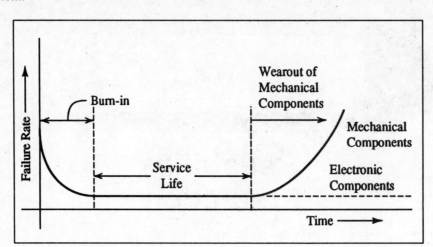

becomes a short or open circuit. Note that variations in component value that are still within manufacturer's specification do not constitute failure.

Figure 4.1 shows idealized failure rate curves for both electronic and mechanical components. Note that the failure characteristic for mechanical parts demonstrates an initial period of high failure rate, followed by a relatively stable rate during the component's service life, followed by a wear-out period in which the failure rate again rises. This pattern of failure rate is called a *bathtub curve*. Electronic components exhibit a similar curve, also shown in Fig. 4.1, except that most of them don't experience the wear-out phase.

The initial high failure rate period is called the *burn-in period;* failures during this period are said to result from "infant mortality." In systems that demand high reliability, manufacturers typically operate the equipment for several hours before delivering it to the user in order to weed out any that will fail for this reason. Infant mortality failures are not covered in the following discussion of reliability.

Most electronic components and systems exhibit constant failure rates during normal product life. *Constant failure rate* means that during any period of time, a fixed percentage of units will fail. For example, suppose 1000 transistor radios were placed in service (after burn-in) and within the first 100 hr, n number of the radios failed. Now, let's replace those failed radios and run the test for another 100 hr. During this latter test period, we would expect n number of the radios to fail; some of these will be from the replacement units, others from the original set. We could repeat this process any number of times with an average of n number of radios failing each test period. The failure rate in this case is expressed as

$$\lambda = \frac{n}{1000} \bullet \frac{1}{100} \text{ failures per hour} \tag{4.1}$$

or, because failure rate is usually expressed in failures per million hours,

$$\lambda = 10n \text{ failures/million hours.}$$

Mathematically, failure rate is easier to express if failed devices are not replaced. The percentage of failures of the remaining devices within any time period remains the same. If n_o devices were placed in service at time $t = 0$, the number remaining at any time t (with no replacement) will follow an exponential decay curve, expressed as follows:

$$n(t) = n_o e^{-\lambda t}. \tag{4.2}$$

As stated earlier, reliability is the probability that a given circuit, device, or system will be operable (has not failed) up to a specified time t. It can be expressed by the following equation:

$$R(t) = e^{-\lambda t}. \tag{4.3}$$

Another useful concept when dealing with reliability is the expected lifetime of a component or system. The expected lifetime is called the *mean-time-between-failures,* or MTBF. The MTBF is the inverse of failure rate and is expressed by the following equation:

$$MTBF = \frac{1}{\lambda}. \tag{4.4}$$

Our discussion so far has dealt with electronic systems with no redundancy. *Redundancy* means that two or more components or subsystems are operated in such a manner that if one fails, the other takes over the function of the failed component or subsystem and the system continues to operate. Redundancy is common in military systems or in transportation systems where a failure could endanger passenger lives. For example, modern automobiles have redundant braking systems, that is, they have two separate master brake cylinders so that if one fails, the driver can stop the car with the other. If a system has redundancy, the calculation of reliability is more complex. We discuss that issue later.

Failure rates for mechanical and electrical devices that experience wear-out and fatigue are not constant and often follow a Weibull, or double exponential, distribution. Because the failure rate, and therefore also the mathematical expression, for mechanical components is more complex than that for the majority of electronic components, we don't discuss mechanical components in this book.

4.3 MILITARY HANDBOOK 217E

As you can see from our discussion so far, finding the failure rate is the key to predicting reliability. Fortunately, many industries as well as the defense establishment have been concerned with reliability for many years and consequently, have systematically recorded failures of many components and devices under various operating conditions. This data has been analyzed and collected into a military handbook known as MIL-HDBK 217E, *Reliability Prediction of Electronic Equipment,* used by United States defense-related industries. This handbook is updated periodically; the suffix "E" denotes the version published in 1986.

MIL-HDBK 217E contains data on failure rates of electronic components in various application environments. These data are based on the constant failure rate model and thus only approximate the observed component failure rate. Also, this predicted failure rate data does not take into account misapplication of the component in the design or mishandling of the component, device, or system during manufacturing or in the field. The data from MIL-HDBK 217E can really only predict the failure rate. The observed failure rate, and hence the reliability, of any particular product can be determined only from its performance in the field.

4.4 COMPONENT FAILURE RATE CALCULATION

To predict a component's failure rate, you need to take into consideration several factors concerning how the component responds to the application and environment, including the application's environment, manufacturing quality, device complexity, and application stress.

For example, the failure rate for a fixed composition resistor is given by the equation

$$\lambda_p = \lambda_b \left(\prod_E \bullet \prod_R \bullet \prod_Q \right) \text{ failures}/10^6 \text{ hr} \tag{4.5}$$

where

λ_p is the component failure rate,

λ_b is the basic failure rate,

$\prod_E$ is the environmental factor,

$\prod_R$ is the resistance factor, and

$\prod_Q$ is the quality factor.

(See Reliability Data in Part 2.)

Note in Table R.12 in Part 2 that the basic failure rate depends on the ambient temperature and on a stress factor that is the ratio of the operating power dissipation to the rated power dissipation.

The environmental factor is given in Part 2 for each type of application. Typical application environments are laboratory use (G_B, Ground Benign), automotive (G_M, Ground Mobile), and helicopter (A_{RW}, Aircraft, Rotary Wing).

Similarly, the resistance factor, Π_R, is determined by the resistance value. For values up to $100 \, k\Omega$, the resistance factor is 1.0; larger resistance values have a higher resistance factor.

The quality factor, Π_Q, also is given in Part 2. Industrial grade has a quality factor of 15, while military grades can be as low as 0.03. A high-quality component, which will have a low failure rate, will have a low quality factor. For example, if you specified a 22 $k\Omega$, 1/2-W, industrial grade, carbon-composition resistor, operating at 1/4 W at 70°C, in a portable radio (Manpack environment, M_P), the predicted failure rate is

$$\lambda_p = 0.0021 \bullet 8.5 \bullet 1.0 \bullet 15 \tag{4.6}$$
$$= 0.26775 \text{ failures}/10^6 \text{ hr},$$

where, from the Tables R.8, 9, 10, and 12 in Part 2,

$\lambda_b = 0.0021$

$\prod_E = 8.5$

$\prod_R = 1.0$

$\prod_Q = 15.$

The failure rate calculation for integrated circuits is more complex. For example, the equation for failure rate of a monolithic (single chip of silicon) digital integrated circuit is given by the equation

$$\lambda_p = \prod_Q (C_1 \bullet \prod_T \bullet \prod_V + C_2 \bullet \prod_E) \prod_L \text{ failures}/10^6 \text{ hr,}$$

where

λ_p is the component failure rate,

$\prod_Q$ is the chip quality factor,

$\prod_T$ is the temperature acceleration factor,

$\prod_V$ is the voltage stress derating factor,

$\prod_E$ is the application environment factor,

$\prod_L$ is the device manufacturer's learning factor,

C_1 is the circuit complexity factor, and

C_2 is the package complexity factor.

The various factors affecting the calculation can be found in Part 2 in tables similar to those given for the resistor. The factors for monolithic digital integrated circuits are given in Tables R.41–46 in Part 2.

4.5 SYSTEM FAILURE RATE CALCULATION

The reliability $R_S(t)$ of a system with several components is the joint probability of all components not failing; that is,

$$R_s(t) = R_1(t) \bullet R_2(t) \bullet R_3(t) \ldots \tag{4.8}$$

When the reliability of each component is expressed in exponential form, then

$$R_s(t) = e^{-\lambda_1 t} \bullet e^{-\lambda_2 t} \bullet e^{-\lambda_3 t} \ldots \tag{4.9}$$

Thus

$$R_s(t) = e^{-(\lambda_1 + \lambda_2 + \lambda_3 + \ldots)t}. \tag{4.10}$$

Therefore the failure rate of the system can be expressed as the sum of the individual component failure rates,

$$\lambda_s = \sum_{i=1}^{n} \lambda_{p,i}, \tag{4.11}$$

where λ_s is the system failure rate and $\lambda_{p,i}$ is the failure rate of the ith component. The components of the system include all electronic components as well as all wiring, connectors, and even the printed circuit boards.

Example 4.1. A Sample Failure Rate Calculation.

Suppose we want to predict the failure rate of an automobile electronic system that consists of a VLSI circuit (very-large-scale integrated circuit), three 1/4-W resistors operating at 1/4 W, and an LED (light emitting diode) used as an indicator light.

The temperature in the automobile may be as high as $70^\circ C$. An automotive application is similar to the military ground mobile—G_M—environment. The failure rate of each resistor is

$$\lambda_p = \lambda_b \, (\prod_E \bullet \prod_R \bullet \prod_Q) \text{ failures}/10^6 \text{ hr}$$

$$= (0.0059) \, (8.3 \bullet 1.0 \bullet 15.0)$$

$$= 0.73455 \text{ failures}/10^6 \text{ hr,}$$

where

$$\lambda_b = .0059,$$

$$\prod_E = 8.3,$$

$$\prod_R = 1.0, \text{ and}$$

$$\prod_Q = 15.0.$$

The failure rate for the integrated circuit is

$$\lambda_p = \prod_Q (C_1 \bullet \prod_T \bullet \prod_v + C_2 \bullet \prod_E) \prod_L \text{ failures}/10^6 \text{ hr}$$

$$= 20.0 \, (\, 0.16 \bullet 7.5 \bullet 1.0 + 0.019 \bullet 4.2 \,) \, 1.0$$

$$= 25.596 \text{ failures}/10^6 \text{ hr,}$$

where

$$\prod_Q = 20,$$

$$C_1 = 0.16,$$

$$\prod_T = 7.5,$$

$$\prod_v = 1.0,$$

$$C_2 = 0.019,$$

$$\prod_E = 4.2, \text{ and}$$

$$\prod_L = 1.0.$$

The failure rate for the LED is

$$\lambda_p = \lambda_b \, (\prod_T \bullet \prod_E \bullet \prod_Q) \text{ failures } 10^6 \text{ hr}$$

$$= 0.00065 \, (\, 1500 \bullet 7.8 \bullet 1.0 \,)$$

$$= 7.605 \text{ failures}/10^6 \text{ hr,}$$

where

$$\lambda_b = 0.00065,$$

$$\Pi_T = 1500,$$

$$\Pi_E = 7.8, \text{and}$$

$$\Pi_Q = 1.0.$$

The predicted failure rate for the complete system is the sum of the failure rates of the components. Thus the system failure rate is predicted to be

$$\lambda_{sys} = \sum_{i=1}^{n} \lambda_{p,i} = 0.73455 + 0.73455 + 25.596 + 7.605$$

$$= 35.40465 \text{ failures}/10^6 \text{ hr}$$

and the MTBF is predicted to be

$$MTBF = \frac{1}{\lambda_{sys}} = \frac{1}{35.40465 \dfrac{\text{failures}}{10^6 \text{ hrs}}} = 28,244.88 \frac{\text{hours}}{\text{failure}}.$$

The reliability prediction is based on the number of hours the system will be in service; for example, an automobile is typically in service for 500 hr/yr. Thus the predicted reliability for 1 yr is

$$R\,(500\,hr) = e^{-[(35.40465 \frac{\text{failures}}{10^6 \text{ hrs}})\,(500\,hr)]} = 0.98253,$$

which means that the probability the system will run for 1 yr without failing is 98% .The probability of failing, $Q(t)$, is

$$Q(t) = 1 - R(t). \tag{4.12}$$

Let's say an automobile could be expected to run for 10 yr. In this case, the predicted reliability over the life of the automobile is

$$R\,(5000\,hr\,) = e^{-[(35.404654 \frac{\text{failures}}{10^6 \text{ hrs}})\,(5000\,hr)]} = 0.8377603$$

$$= 84\%.$$

If the system in Example 4.1 were to operate continuously, for example, as a clock or a burglar alarm does, the reliability would be based on 8760 hr/yr (365 • 24 = 8760). In this case, the reliability for 1 yr would be

$$R(8760\,hr) = e^{-\left[(35.40465 \frac{\text{failures}}{10^6 \text{ hrs}})\,(8760\,hr)\right]} = 0.7333408.$$

$$= 73\%.$$

If several of these units were placed in service, more than 1/4 would be expected to fail within 1 yr.

For 10 yr, the predicted reliability would be

$$R(87600\,hr) = e^{-\left[(35.40465 \frac{\text{failures}}{10^6 \text{ hrs}})\,(87,600\,hr)\right]} = 0.04498405$$

$$= 4.5\%.$$

It would be unlikely that the example system would run continuously for 10 yr without failure.

4.6 SYSTEMS WITH REDUNDANCY

When two components are placed in parallel redundancy, the resulting system failure rate is lower than it is for the individual components. This is because both components must fail before the system fails. In this case, the probability that both fail is

$$Q_R(t) = Q_1(t) \ Q_2(t), \tag{4.13}$$

where

$$Q_1(t) = 1 - R_1(t) \text{ and } Q_2(t) = 1 - R_2(t).$$

Thus the reliability of the redundant combination is

$$R_R(t) = 1 - Q_R(t)$$

$$= 1 - \{[\ 1 - R_1(t)] \bullet [1 - R_2(t)]\}$$

$$= R_1(t) + R_2(t) - R_1(t) \bullet R_2(t). \tag{4.14}$$

Now, if we write the reliability of the redundant combination as

$$R_R(t) = e^{-\lambda_r t}, \tag{4.15}$$

then λ_R can be found by taking the logarithm,

$$\lambda_R = \ln \left[e^{-\lambda_1 t} \bullet e^{-\lambda_2 t} \bullet e^{-(\lambda_1 + \lambda_2)t} \right]. \tag{4.16}$$

When a redundant section is part of a larger system, the total system reliability is the product of the subsystem or component reliabilities, as follows:

$$R(t) = R_R(t) \bullet R_N(t) \tag{4.17}$$

where $R_R(t)$ is the reliability of the redundant section and $R_N(t)$ is the reliability of the rest of the system.

The reliability prediction of more complex systems having multiple redundant subsystems is accomplished by finding the reliability predictions for each subsystem and then combining those predictions. More details are given in the references.

4.7 BIBLIOGRAPHY

1. Dhillon, B.S. *Reliability Engineering in System Design and Operation*. New York: Van Nostrand-Reinhold, 1983.

2. Shooman, M.L. *Probabilistic Reliability: An Engineering Approach*. New York: McGraw-Hill, 1968.

5

Designing for Safety

"Carriages without horses shall go,
And accidents fill the world with woe."

Prophecy attributed to Mother Shipton
[17th century]

5.1 INTRODUCTION

Engineers have an obligation to society. No design of a device that could possibly cause harm can be considered complete until the designer has taken all reasonable precautions to eliminate or mitigate that possibility. Should a device cause injury or property damage that results from foreseeable equipment failure or misuse by its operator, then the designer could be held negligent.

Well-designed systems often will work satisfactorily, that is, within specifications, for a long time. Such designs are considered to be reliable. Eventually, however, all systems fail, usually because a single component has ceased to work according to specifications; this failure can cause a chain-reaction of other component failures. We discuss design reliability in Chapter 4. However, it's not enough that a system be reliable. Because we can expect failure of a system at one time or another, it's important that such failure cause as little harm as possible. The principal design evaluation procedure we use to uncover unsafe failures is the Failure Mode and Effects Analysis (FMEA).

5.2 FAILURE MODE AND EFFECTS ANALYSIS

FMEA provides a means of systematically evaluating a design for possible adverse effects and involves the following five steps:

1. Systematically label all components on the schematic diagram.
2. State concisely the function of each component within the system.
3. Identify the failure modes for each component.

4. Determine the effect of each single failure on system operation. In some cases a single component failure will cause a chain-reaction series of failures, known collectively as a **common-mode failure**; your analysis must account for such failures.
5. Identify those failures that could cause actual or potential harm or operational failure.

After completing the FMEA, you can modify the design to alleviate the most harmful actual or potential effects. To do this, you must ensure not only that the design is complete, but also that all components are included, especially easy-to-overlook ones such as interconnecting wires or connectors. Therefore, to help ensure FMEA is performed thoroughly, that is, that no component is overlooked, usually you will use special forms. Let's consider an example to demonstrate how the FMEA is performed.

Example 5.1. A simple traffic signal.

Figure 5.1 shows a schematic diagram for a simple traffic signal. The signal display consists of a single four-sided unit suspended over the center of a right-angle crossing intersection. There are 12 signal lamps: a red one, yellow one, and green one for each of the four directions. Opposing signal lamps are powered in parallel by a single circuit. There are six power lines and a single common return line for the display for a total of seven wires. A seven-wire cable runs from a control unit to the display. The control unit is mounted on a pole at one corner and consists of a small electric motor and gear box turning a cam shaft at 1 rpm. The cams on the shaft close switches at appropriate times to light the 12 lamps in the desired sequence. The system runs directly from 120 V, 60 Hz power. The traffic signal display sequence can be altered only by replacing the cam shaft with a different one. The signal does not respond to traffic conditions or provide separate signals for left turns.

The FMEA is performed as follows:

Step 1. Systematically list all components on a schematic diagram.

When listing these components, include unique identifying codes, such as R1, R2, J1, etc. It's important that this step be done thoroughly. The labels are shown in place in Fig. 5.1. Note that we have labeled the interconnecting wires as well as all the components such as the lamps, switches, and motor.

Step 2. State concisely the function of each component within the system.

In Example 5.1, the results of step 2 are shown in the second column of Fig. 5.3. All the components from Fig. 5.1 are listed in the first column (notice we made certain to list everything). The function of NR is to display a "stop" (Red) signal to the North, etc.

Step 3. Identify failure modes for each component.

For electrical and electronic systems, the components are usually basic circuit elements; therefore the component failure modes are limited. For example, a resistor can either fail open circuit or fail short circuit, a logic gate can get stuck at "1" or stuck at "0", or a wire can break off or become shorted to another wire. Often FMEA concentrates on the limiting situations where a component fails completely, but sometimes it is important to consider degradation failures, where the component falls outside specification but does not fail completely.

Figure 5.1 Traffic Signal Schematic Diagram (inset shows operation of cam switches).

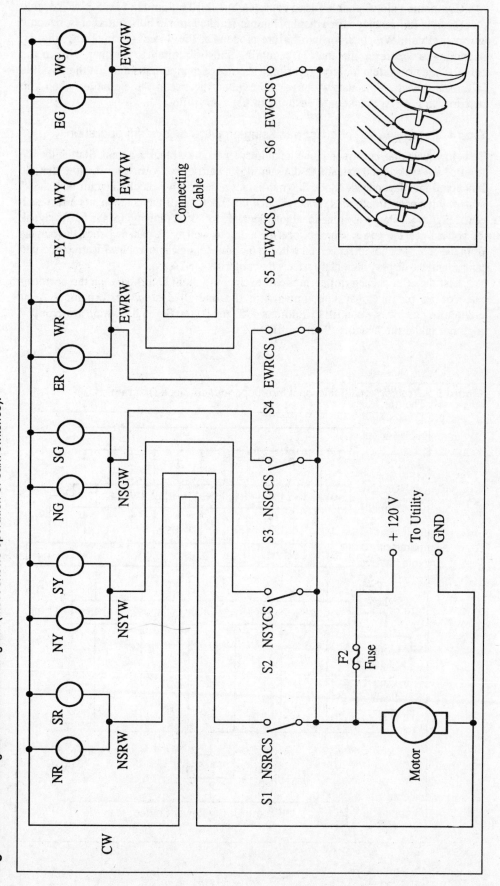

Completing step 3 enables us to complete the third column in Fig. 5.3. We decide that the only failure mode for a display lamp is for the lamp to burn out, that is, become an open circuit. We consider two failure modes for the wires: (1) breaking, that is, becoming an open circuit and (2) becoming short-circuited to another wire in the cable. Then we systematically go through the list of components and list the possible failures for each. Note that we have abbreviated the list in the example; when you perform an FMEA, you should ensure you list everything.

Step 4. Determine the effect of each single failure on system operation.

To help you do this, construct a functional interdependence block diagram. Start at the top level of the system and extend the diagram downward to the lowest level. Note that the functional interdependence block diagram is not a descriptive block diagram that shows system interconnection; rather, it shows how the functions of the components affect each other. Figure 5.2 shows the functional interdependence block diagram for the traffic signal. As you can see, the power source affects the motor as well as the cam switches, the outputs from the cam switches affect certain wires in the connecting cable, each wire affects certain lamps, and the display ultimately affects the motorists who see it.

List the effects each failure mode of each component could have on the performance of its parent subsystem, upward to the overall system performance. After completing step 4, we can fill in columns 4, 5, and 6 (see Fig. 5.3), which are grouped together under the heading "Failure Effect."

Figure 5.2. Traffic Signal Functional Interdependence Block Diagram.

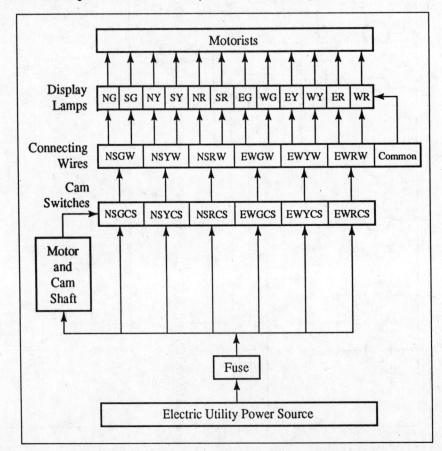

Figure 5.3. Example Failure Mode and Effects Format Sheet for Traffic Light.

Part Designation	Function	Electrical Failure Mode	Failure Effect				Possible Design Improvement	Remarks
			Component	Sub System	Overall System*			
Lamp-north facing red (NR)	North-stop	Open	Doesn't Light	No NR Display	Moderate Traffic Hazard	Redundant Lamps	Dual Lamps Cheap
Lamp NY	North-caution	Open	Doesn't Light	No NG Display	No Traffic Hazard	None Required	
Lamp NG	North-go	Open	Doesn't Light	No NG Display	No Traffic Hazard	None Required	
9 more lamps failed open							
Wire CW	Common Return	Open	Won't Carry Current	No Displays	Moderate Traffic Hazard	Individual Returns For All Circuits	
Wire NSGW	North-go South-go	Open	Won't Carry Current	No NG, SG Displays	Moderate Traffic Hazard	Redundant Wires	
		Short to EWGW	Improperly Energized	Simultaneous NG, SG, EG,WG, NR, SR Display	Significant Traffic Hazard**	Special Wiring System Monitor	What is the Probability?

Figure 5.3. Example Failure Mode and Effects Format Sheet for Traffic Lght. (Continued)

Part Designation	Function	Electrical Failure Mode	Failure Effect			Possible Design Improvement	Remarks
			Component	Sub System	Overall System*		
5 more wires failed open							
20 more 2-wire short circuit failures							
Cam Switch NSGCS	North-go South-go	Open	Won't Carry Current	No NG, SG Display	Moderate Traffic Hazard	Redundant Switches	
		Short	Always Closed	NG, SG Always On	Significant Traffic Hazard	Series Switches, System Monitor	Monitor Logic ?
5 other switches failed open and failed short							
Drive Motor	Turn Shaft	Burn Up	Stop	Stuck Display	Low Traffic Hazard**	None Required	
Power Supply	Energize	Off	No Power	No Display	Moderate Traffic Hazard	Emergency Power Supply	Not Feasible
Fuse	Circuit Protection	Open	No Power	No Display	Moderate Traffic Hazard	Slow-blow	

* Assume that overall system comprises the signal and the interacting motorists.

** How drivers react to certain display situations may require further study.

Step 5. Identify those failures that could cause actual or potential harm or operational failure.

Once you have completed the FMEA, you should have the data you need to determine actual or potential harm that could result from component failure in the circuit, device, or system you have designed. In Example 5.1, we now know that a significant traffic hazard exists for those failures that would cause the signal to display green simultaneously in all directions. As a designer, you next must modify the design so that the cause of this failure is reduced or eliminated. The last two columns show our response to this hazard and lists possible design modifications.

5.3 DESIGN MODIFICATIONS FOR SAFETY

Once you have identified failure modes that have actual or potential undesirable effects, you can choose from four types of design modifications to reduce or eliminate the effects, as follows:

1. Eliminate the possibility of a failure by eliminating the particular component failure mode causing the problem.
2. Increase the reliability of critical components to reduce the likelihood of a harmful failure.
3. Redesign the system to be tolerant of the particular component failure through the use of the principle of redundancy.
4. Design a monitoring system to provide a warning or protection in the event of primary system failure.

Generally, the most desirable design modification is to eliminate the possibility of a failure by eliminating the particular component failure mode causing the problem. This approach is the most desirable because it doesn't increase the complexity of your circuit or system or affect its reliability. For example, a smoke detector that fails by setting off an alarm when no fire is present is preferred to one that fails to set off an alarm when a fire does occur. In Example 5.1, a traffic signal failure in which the signal displays red to all traffic is much preferred over a failure in which green is displayed to all traffic or one in which the traffic signal is simply turned off. Often, failure effects can be changed from potentially catastrophic to safe by such simple measures as using "active low" configurations instead of "active high" configurations or by detecting the absence of a signal instead of its presence.

A second alternative is to increase the reliability of critical components so that a harmful failure is less likely. We discuss reliability in Chapter 4. Note that this approach doesn't affect the consequences of a failure, but only the probability that it will occur.

Methods three and four, while effective, require great care be taken in their implementation. These alternatives are less desirable than are the first two methods discussed. They are "add-on" design modifications that result in increased system complexity and reduced reliability, and consequently, higher costs (both manufacturing and maintenance). Additionally, in either case, a way must be designed to reliably detect either that portions of a redundant system have failed or that a monitoring system has failed. This disadvantage is further exacerbated when the monitor or redundant part fails and the primary system then reverts to being subject to an unsafe failure—a situation that must be detected and corrected as soon as possible.

Detecting the failure of a redundant part or a monitoring system usually can be accomplished only by checking periodically to see if the part or system still works. Some sophisticated systems can run self-tests either periodically or at system start-up. You would need to design the test carefully to ensure it checks the entire redundant component or monitoring system. An example of such a test is a "push-to-test" button, which is a common method to manually determine if a monitoring system is operational. The best test would be one that most nearly represents the conditions the monitor is designed to detect, such as actually blowing smoke on a home smoke detector.

5.4 BIBLIOGRAPHY

1. Barbour, G.L. "Failure Modes and Effects Analysis by Matrix Method." Proceedings 1977 Annual Reliability and Maintainability Symposium, IEEE 77CH1161-9RQC. Philadelphia, PA. 1977. p. 114-119.

2. Failure Mode/Mechanism Distributions 1991, FMD 1991. Reliability Analysis Center. IIT Research Institute. Rome, NY. 1991.

3. MB-350-B: A Guideline for the FMEA/FTA. Westinghouse Corporate Product Integrity. Westinghouse Research and Development Center. Churchill, PA. 1981. 8pp.

6

Design Optimization

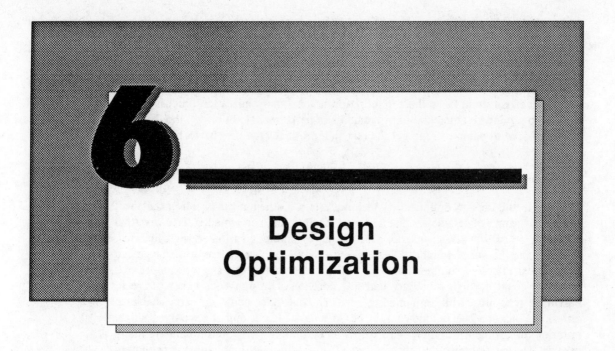

*"... all parts of the system must be constructed
with reference to all other parts, since,
in one sense, all the parts form one machine,
and the connections between the parts being
electrical instead of mechanical. Like any
other machine the failure of one part to
cooperate properly with the other part
disorganizes the whole and renders it
inoperative for the purpose intended."*

Thomas Edison

6.1 INTRODUCTION

The design process involves making decisions. For example, the designer decides which values of resistors to use in a circuit, which diameter bolts to use in a joint, or which diameter wire to use for a power distribution system. Variables such as these that are under the designer's control are the *parameters* of the design. *Design optimization* is the process of determining which choices of design parameters will produce the best, that is, the most desirable, design.

Developing an optimum design is a very different activity from finding the correct answer to a typical engineering homework problem. No design is likely to be perfect—some are simply better or worse than others. Many designs may work satisfactorily; by optimizing your design, you strive to produce the best possible result.

6.2 TWO STAGES OF OPTIMIZATION

Design optimization usually consists of two stages. First, the designer considers possible general design configurations. These configurations are often distinct and mutually exclusive, such as using either a series versus a parallel circuit or an infrared versus a radio transmission link. The process by which one basic design configuration is chosen over another varies widely; the decision could be based on the personal preference of the designer or on a sophisticated performance matrix analysis. We will not attempt in this book to deal with this stage of design decision making.

After the basic configuration of the design is chosen, the designer selects the numerical values of the components that will produce acceptable overall performance for that configuration. From there, a more formal design optimization process can be applied to produce incremental improvements in performance. In effect, the designer "tweaks" several parameters to get the mix that produces the best performance.

6.3 CRITERION FUNCTION

The key to this design optimization technique is a mathematical equation that best expresses the mix of parameters that will produce optimum performance. To derive this equation, called a *criterion function,* the designer combines various design features, such as cost, size, speed, reliability, safety, etc., making adjustments as needed to arrive at the most effective combination based on the design objective. For example, the objective could be high reliability and low cost. Your design would then strive to maximize reliability while minimizing cost and this ratio would be expressed as a criterion function. The function might be so simple that it embodies only a single criterion or sufficiently complex to express a complex interrelationship among many criteria. Design optimization, then, consists of finding the maximum or minimum possible value for the criterion function so that the resulting design has an optimum blend of the desirable criteria. Example 2.2 demonstrates the solution to a design optimization problem. In Example, 6.1, we explore the use of criterion functions in design optimization.

Design optimization consists of the following steps:

1. Determine the features that are important to evaluating the quality of the design.
2. Express the importance of the criteria relative to each other.
3. Develop analytical relationships between the parameters of the design and the terms in the criterion function.
4. Combine the relationships to yield a single function that quantitatively expresses the overall performance of the design.
5. Determine values of design parameters that will maximize the criterion function.

Example 6.1. Design optimization.

Your company has been contracted to provide dc power to a fixed load ($10\ \Omega$) continuously for 10,000 hr. The customer wants to pass 50–75 A through the load and will pay 10¢/kWh for the power. Your company has a 1000-V dc source available 3 km from the customer's load. In order to supply power, your company must build a line to the customer, deliver the power, and then dismantle the line at the end of 10,000 hr (see Fig. 6.1).

Your company estimates that the cost of constructing, maintaining, and dismantling the line itself is directly proportional to the weight of copper in the wire used at $18/kg. Further, it is estimated that the company's cost to deliver energy to the customer (in addition to line cost) is 3¢/kWh produced by the dc source. You are to choose the wire size to use in building the line that will provide your company with the largest percentage profit. What size copper conductor should you choose?

Step 1. Determine the features that are important in evaluating the quality of a design.

In comparing one design to another, what features should be considered? Refer back to the example narrative to see that profit is the quantity you want to optimize. This criterion function can be expressed mathematically as

Figure 6.1. Design Optimzation Example–Power Line Sizing.

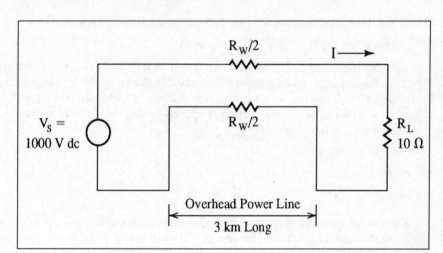

$$\text{Profit (P)} = \frac{\text{revenue (r)} - \text{costs (c)}}{\text{costs (c)}} \cdot 100\% \tag{6.1}$$

P is to be made as large as possible.

We can see from Eq. 6.1 that we will need to determine quantitatively both the revenue and the cost of delivering power.

Step 2. Express quantitatively the importance of the criteria relative to each other.

Assign coefficients in the criterion function to indicate this importance. In the example, we express only a single criterion—profit—so we can skip step 3. However, if we needed to consider the reliability of the transmission line as well as its cost, then we would need to decide on the relative importance of saving d dollars versus increasing reliability by x percent. If we could develop a functional relationship between cost and reliability, we could then compute reliability in terms of dollars in order to determine the coefficients; otherwise, we would need to make a subjective judgment to choose the coefficients.

Step 3. Develop analytical relationships between the parameters of the design and the terms in the criterion function.

We need to compute both revenue and cost. Revenue is 10¢/kWh delivered to the load R_L (10 Ω). The power delivered is $I^2 R_L$. The following equation applies:

Revenue(r) = rate • time • power, or

$$r = \left[\frac{(0.1\ \$/\text{kWh})}{1000} (10{,}000\,\text{hr})(10\,\Omega) \right] I^2.$$

$$r\ (\text{in } \$) = 10\,I^2\ (\text{in A}^2). \tag{6.2}$$

Costs are \$18/kg for wire plus 3¢/kWh for energy produced by the source, as expressed by the following equation:

Costs (c) = (power generation costs) + (capital costs), or

c = (rate • time • power) + (line installaton costs), or

$$c = \left[\frac{(0.03\$/kWh)}{1000}(10,000\,hr)(1000\,V)\right]I + (18\,Wt)$$

Where Wt is the weight of the wire in the lines in kilograms. Thus

$$c\text{ (in \$)} = (300\,I) + (18\,Wt). \tag{6.3}$$

Wire is available only in standard sizes, so the relationship between wire size and capital cost is a set of discrete points. The smaller the wire, the less expensive it is. On the other hand, smaller wire has higher resistance, which both increases the losses during power transmission and decreases the power delivered to the customer.

The current in the circuit is

$$I = (V_S)/(R_W + R_L), \tag{6.4}$$

where

$$V_S = 1000\,V,\ R_L = 10\,\Omega.$$

The capital cost of the transmission line and its resistance are not expressed by simple mathematical functions, however. They are found in tables of data for each size wire. Table 6.1 shows this information.

Step 4. Combine the relationships to yield a single function (the criterion function) that expresses quantitatively the overall performance of the design.

In the example, we need to combine Eqs. 6.1, 6.2, 6.3, and 6.4. Using Eq. 6.4,

$$I = \frac{1000}{10 + R_W}.$$

Plugging this equation into Eqs. 6.3 and 6.2, we arrive at

$$c = \frac{300 \cdot 1000}{10 + R_W} + 18\,Wt$$

and

$$r = 10\left[\frac{1000}{10 + R_W}\right]^2.$$

Table 6.1. Line weight (Wt), line resistance (R_W).

Wire Size (AWG)	Weight/6 km (Wt)	Resistance/6 km (R_W)
1	2216 kg	2.54 Ω
2	1758	3.20
3	1394	4.03
4	1105	5.09
5	877	6.42
6	695	8.20
7	551	10.20
8	437	12.84
9	346	16.20

These equations can be substituted into Eq. 6.1 as follows:

$$P = \left\{ \frac{10\left[\frac{1000}{10+Rw}\right]^2 - \left[\frac{300 \cdot 1000}{10+Rw} + 18\,Wt\right]}{\left[\frac{300 \cdot 1000}{10+Rw} + 18\,Wt\right]} \right\} 100\% \qquad (6.5)$$

Step 5. Determine values of design parameters that will maximize the criterion function.

We discuss this step in the next section.

6.4 OPTIMIZATION TECHNIQUES

There are several ways to determine the values of the design parameters to maximize the criterion function. One is to try various values of the parameters in the equation, searching for the combination of values that gives the maximum value for the criterion function. Although straightforward, this approach might not be particularly efficient and so should be reserved for cases in which the criterion function is too complex to evaluate any other way.

A second method involves plotting the value of the criterion function versus the design parameter and then searching for the maximum. This works for criterion functions incorporating one or two design parameters but not for more complex relationships.

A third method is to determine the optimum values of the parameter analytically by setting the derivatives of the criterion function equal to zero. If the function depends on more than one design parameter, successive partial derivatives must be taken with respect to each design parameter. Doing this produces a set of simultaneous equations that can then be solved for the optimum values of the design parameters.

We will implement step 5 using both the method of direct substitution and that of plotting the criterion function versus the design parameter (in this case, the wire size). The variables Rw and Wt in Eq. 6.5 are discrete values taken from Table 6.1. By evaluating Eq. 6.5 for each line of Table 6.1, we can produce Table 6.2. Consequently, we can readily see that AWG #4 wire maximizes the criterion function. Note that

Table 6.2. Criterion function vs. wire size.

Wire Size (AWG)	Total Costs(c)	Revenue(r)	Criterion Function(P)
1	$63,812	$63,585	- 0.36
2	54,366	57,396	+ 5.57
3	46,471	50,808	+ 9.33
4	39,771	43,917	+10.42
5	34,050	37,088	+ 8.92
6	28,989	30,195	+ 4.16
7	24,774	24,512	- 1.06
8	21,008	19,167	- 8.76
9	17,687	14,569	- 17.63

performing these activities represents an ideal application of a computer spreadsheet program (as was the design optimization in Example 2.2). We encourage you to use spreadsheets whenever possible to optimize designs.

If we plot the criterion function versus wire size, we get the plot shown in Fig. 6.2. Notice that the optimum wire size is AWG #4.

As you tackle more complex designs, the design optimization procedure becomes much more difficult, for the following reasons:

1. The number of criteria to be incorporated into the criterion function increases. For example, in designing a stereo amplifier, the power output, frequency response, distortion, size, price, reliability, and many other characteristics of the amplifier are probably all important to users and therefore deserve consideration.

2. The relative importance of various characteristics might be difficult to determine. For example, the designer might not know the relationship between the power output of an amplifier and how much customers are willing to pay. Or the designer might not anticipate that improvement in materials will radically change the relationship between size and cost. However, the weighting factors (coefficients) assigned to the various criteria are critical in determining the overall design; if they can't be measured or computed, they must be estimated.

3. The mathematical complexity involved in finding an optimum increases very rapidly as the number of design parameters increases; consequently,

Figure 6.2. Criterion Function vs. Wire Size to Optimize Power Line Design.

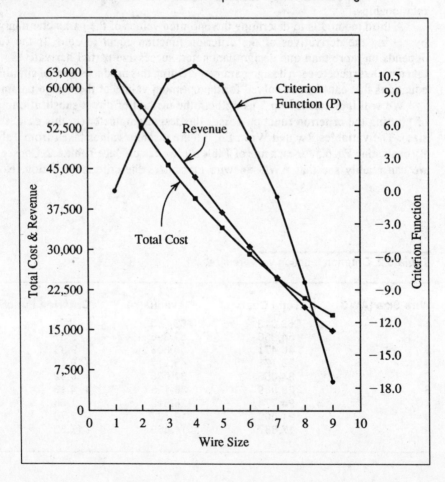

exhaustive computer-based searches become necessary. Sophisticated search techniques help somewhat, but many complex optimization problems have not yet been completely explored.

Because of these difficulties, many important design decisions must be made without a rigorous formal design optimization. How good the resulting design proves to be hinges on the experience and intuition of the designer and on the designer's ability to make effective decisions without sufficient data to provide adequate guidance.

As a student designer, you easily could be overwhelmed by a complex optimization task. In this book, we assist you by partially or completely specifying the information you need to develop the criterion function, so that your major efforts can be concentrated on developing relationships between design parameters and the criteria expressed by the criterion function.

6.5 BIBLIOGRAPHY

1. Caney, Steven. *Steven Caney's Invention Book*. New York: Workman Publishing, 1985. 207 pp.

2. Middendorf, W.H. *Design of Devices and Systems*. New York: Marcel Dekker, Inc., 1986, pp 229-297, 374-406.

Part II
Data Sheets

Approved Parts

Cost, Specifications, and Reliability

We obtained cost information from many sources. Therefore, the prices given reflect reasonable values at the time these lists were assembled. Because prices often change, prices given might bear little resemblance to present market prices.

Component specifications were obtained from manufactures' data sheets and represent only a small number of those actually available on the market. In many cases, we condensed the data, but to the extent possible, we retained their accuracy. However, actual specifications can vary somewhat from those given herein.

Passive Components

COPPER WIRE TABLES

Standard Annealed Copper,   American wire gage

	20°C			Resistance Variation with Temperature			
Gage AWG	Diameter (mm)	Ohms per kilometer	Kg per kilometer	Ohms per 1000 feet 0°C	20°C	50°C	100°C
0000	11.68	0.1608	953.2	0.0452	0.0490	0.0548	0.0644
000	10.40	.2028	755.8	.0570	.0618	.0691	.0813
00	9.266	.2557	599.5	.0718	.0779	.0871	.1024
0	8.252	.3223	475.5	.0905	.0982	.1098	.1291
1	7.348	.4065	377.0	.1142	.1239	.1385	.1629
2	6.543	.5128	298.9	.1440	.1563	.1747	.2054
3	5.827	.6466	237.1	.1816	.1971	.2203	.2590
4	5.189	.8152	188.0	.2289	.2485	.2778	.3266
5	4.620	1.028	149.0	.2888	.3134	.3504	.4120
6	4.115	1.297	118.2	.3641	.3952	.4418	.5194
7	3.665	1.634	93.80	.4589	.4981	.5568	.6547
8	3.264	2.061	74.38	.5787	.6281	.7021	.8256
9	2.906	2.600	58.95	.7302	.7925	.8859	1.042
10	2.588	3.277	46.77	.9203	.9988	1.117	1.313
11	2.30	4.14	37.1	1.16	1.26	1.41	1.66
12	2.05	5.21	29.4	1.46	1.59	1.78	2.09
13	1.83	6.56	23.4	1.84	2.00	2.24	2.63
14	1.63	8.28	18.5	2.33	2.52	2.82	3.32
15	1.45	10.4	14.7	2.93	3.18	3.56	4.18
16	1.29	13.2	11.6	3.70	4.02	4.49	5.28
17	1.15	16.6	9.24	4.66	5.05	5.65	6.64
18	1.02	21.0	7.32	5.88	6.39	7.14	8.39
19	0.912	26.4	5.81	7.41	8.05	9.00	10.6
20	.813	33.2	4.61	9.33	10.1	11.3	13.3
21	.724	41.9	3.66	11.8	12.8	14.3	16.8
22	.643	53.2	2.88	14.9	16.2	18.1	21.3
23	.574	66.6	2.30	18.7	20.3	22.7	26.7
24	.511	84.2	1.82	23.7	25.7	28.7	33.7
25	.455	106	1.44	29.8	32.4	36.2	42.5
26	.404	135	1.14	37.8	41.0	45.9	53.9
27	.361	169	0.908	47.4	51.4	57.5	67.6
28	.320	214	.715	60.2	65.3	73.0	85.9
29	.287	266	.575	74.8	81.2	90.8	107
30	.254	340	.450	95.6	104	116	136
31	.226	430	.357	121	131	146	172
32	.203	532	.288	149	162	181	213
33	.180	675	.227	190	206	230	270
34	.160	857	.179	241	261	292	343
35	.142	1090	.141	305	331	370	435
36	.127	1360	.113	382	415	464	545
37	.114	1680	.0912	472	512	573	673
38	.102	2130	.0721	597	648	725	852
39	.089	2780	.0552	780	847	946	1110
40	.079	3540	.0433	994	1080	1210	1420

SWITCHES

General purpose *toggle switches*:

Action	Rating	Cost
SPST	125 Vac	1.69
SPDT	15 A	2.07
DPST		2.68
DPDT	or	3.20
3PST	250 Vac	5.25
3PDT	10 A	6.46
SPST	125 Vac	1.54
SPDT	or	1.98
DPST	28 Vdc	2.18
DPDT	3A	2.59

General purpose *pushbutton switches*:

Action	Rating	Cost
SPNO	125 Vac	2.21
SPNC	6 A	2.21
SPDT		2.38
SPNO	115 Vac	1.26
SPNC	1 A	1.26
SPNO	50 Vdc	.85
SPNC	25 mA	.85
SPDT		.89

DIP Switches
    Contact ratings: 50 mA @ 24 Vdc
    Life: 7,000 switching cycles
    Switching action: SPST

# of circuits	Unit Cost
2	1.17
3	1.22
4	1.25
5	1.27
6	1.32
7	1.35
8	1.37
9	1.42
10	1.47
12	1.57

Rotary Switches
    Contact ratings: make/break 200 mA
                      carry 6 A cont.
    Life: 25,000 cycles

# poles	# pos.	Unit Cost
1	12	4.77
2	6	5.30

FUSES

10,000 A interrupting capacity @ rated voltage.

Time Delay - 3AG size

Current Rating	Voltage Rating	Unit Cost
1/100 1/32	125	1.67
1/16 1/10 1/8 15/100 .175 3/16 2/10	125	1.32
1/4 3/10 3/8 4/10 1/2 6/10 7/10 3/4 8/10 1	125	.81
1.25 1.5 1.6 2 2.25 2.5 3 3.2 4 5 6.25 7	125	.58
8 10 12 15 20 25 30	32	.78

Fast Acting - 3AG size

Current Rating	Voltage Rating	Unit Cost
1/16 1/8 .175 3/16	125	.84
1/4 3/10 3/8	125	.52
1/2 3/4	125	.36
1 1.5 2 2.5 3	125	.23
4 5 6	125	.35
7 8 10	125	.35 .48 .48
15 20 25 30	32	.23 .19 .23 .23

Fuse Clearing Time

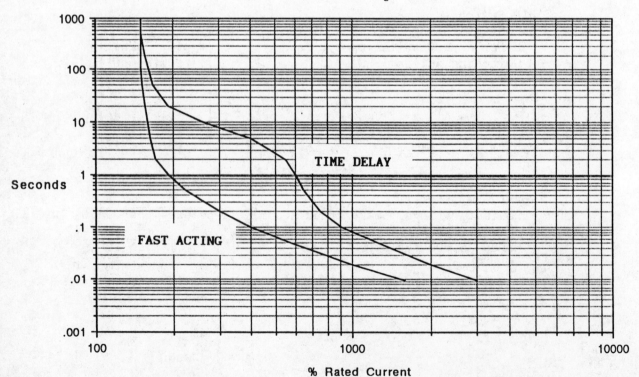

FUSES

10,000 A interrupting capacity @ rated voltage.

Time Delay - 3AG size

Current Rating	Voltage Rating	Unit Cost
1/100 1/32	125	1.67
1/16 1/10 1/8 15/100 .175 3/16 2/10	125	1.32
1/4 3/10 3/8 4/10 1/2 6/10 7/10 3/4 8/10 1	125	.81
1.25 1.5 1.6 2 2.25 2.5 3 3.2 4 5 6.25 7	125	.58
8 10 12 15 20 25 30	32	.78

Fast Acting - 3AG size

Current Rating	Voltage Rating	Unit Cost
1/16 1/8 .175 3/16	125	.84
1/4 3/10 3/8	125	.52
1/2 3/4	125	.36
1 1.5 2 2.5 3	125	.23
4 5 6	125	.35
7 8 10	125	.35 .48 .48
15 20 25 30	32	.23 .19 .23 .23

Fuse Clearing Time

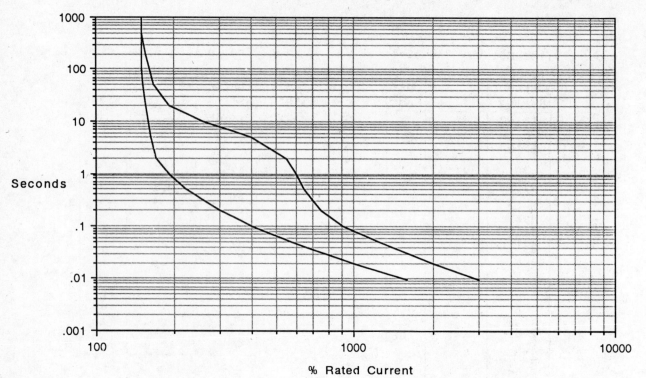

% Rated Current

RESISTORS

Standard Value Sequences for Metal Film (1% Tolerance) and Carbon Resistors (5% and 10% Tolerance).

1%	5%	10%
10.0	10	10
10.2		
10.5		
10.7		
11.0	11	
11.3		
11.5		
11.8	12	12
12.1		
12.4		
12.7		
13.0	13	
13.3		
13.7		
14.0		
14.3		
14.7		
15.0	15	15
15.4		
15.8	16	
16.2		
16.5		
16.9		
17.4		
17.8	18	18
18.2		
18.7		
19.1		
19.6		

1%	5%	10%
20.0	20	
20.5		
21.0		
21.5	22	22
22.1		
22.6		
23.2		
23.7	24	
24.3		
24.9		
25.5		
26.1		
26.7	27	27
27.4		
28.0		
28.7		
29.4	30	
30.1		
30.9		
31.6		
32.4	33	33
33.2		
34.0		
34.8		
35.7	36	
36.5		
37.4		
38.3	39	39
39.2		

1%	5%	10%
40.2		
41.2		
42.2	43	
43.2		
44.2		
45.3		
46.4	47	47
47.5		
48.7		
49.9	51	
51.1		
52.3		
53.6		
54.9	56	56
56.2		
57.6		
59.0		
60.4		

1%	5%	10%
61.9	62	
63.4		
64.9		
66.5	68	68
68.1		
69.8		
71.5		
73.2		
75.0	75	
76.8		
78.7		
80.6	82	82
82.5		
84.5		
86.6		
88.7		
90.9	91	
93.1		
95.3		
97.6		

AVAILABLE VALUES:

Carbon: $1\Omega \le R \le 100\ M\Omega$

Metal Film: $10\Omega \le R \le 10\ M\Omega$

Available values are obtained by multiplying sequence number by a power of 10 (10^{-1}, 10^{0}, 10^{1}, 10^{2}, etc.)

Standard values for wire-wound resistors (5% Tolerance)

0.008	0.5	6	22.5	56	150	400	1 k	4k
.01	.75	7	25	60	160	430	1.2k	5k
.02	1	7.5	27	62	180	450	1.3k	10k
.03	1.5	8	30	70	200	470	1.5k	15k
.05	2	10	33	75	220	500	1.8k	20k
.1	2.5	12	35	80	250	560	2 k	25k
.15	3	15	40	82	270	600	2.2k	40k
.2	3.3	16	45	100	300	680	2.5k	50k
.26	4	20	47	110	330	700	3 k	100k
.3	5	22	50	120	390	750	3.5k	150k
						910		

The Standard Color Code for Resistors

Carbon   Metal Film (1%)

1st Digit	2nd Digit	Multiplier	Tolerance	COLOR	1st Digit	2nd Digit	3rd Digit	Multiplier	Tolerance
0	0	10^0 = 1	–	black	0	0	0	10^0 = 1	–
1	1	10^1 = 10	–	brown	1	1	1	10^1 = 10	± 1%
2	2	10^2 = 100	–	red	2	2	2	10^2 = 100	–
3	3	10^3 = 1k	–	orange	3	3	3	10^3 = 1k	–
4	4	10^4 = 10k	–	yellow	4	4	4	10^4 = 10k	–
5	5	10^5 = 100k	–	green	5	5	5	10^5 = 100k	–
6	6	10^6 = 1M	–	blue	6	6	6	10^6 = 1M	–
7	7	10^7 = 10M	–	violet	7	7	7	10^7 = 10M	–
8	8		–	grey	8	8	8		–
9	9		–	white	9	9	9		–
–	–		± 5%	gold	–	–	–		–
–	–		± 10%	silver	–	–	–		–
–	–		± 20%	no band	–	–	–		–

Unit cost of resistors

Type	Tolerance	Power Rating	Cost
Carbon	5%	1/8 W	0.19
		1/4 W	.08
		1/2 W	.09
		1 W	.14
		2 W	.29
	10%	1/8 W	0.16
		1/4 W	.05
		1/2 W	.06
		1 W	.11
		2 W	.23
Metal Film	1%	1/8 W	0.34
		1/4 W	.60
		1/2 W	1.05
Wire-wound	5%	5 W	1.05
		12 W	2.32
		25 W	3.03
		50 W	4.68
		100 W	7.84
		225 W	13.36

Resistor Temperature Characteristics

Carbon Resistors

Ranges of percent deviation from 25 $^\circ$C resistance value at selected temperatures

Nominal Resistance at 25 $^\circ$C	Resistance Deviation in %								
	-55°C	-25°C	-15°C	0°C	25°C	55°C	65°C	85°C	105°C
≤ 100 kΩ	+0.4 to +7.7	-0.5 to +4.0	-0.6 to +3.0	-0.6 to +1.7	0	-1.2 to +1.5	-1.4 to +2.2	-1.5 to +3.9	-1.2 to +6.0
> 100 kΩ	+0.5 to +10.2	-0.7 to +5.4	-0.9 to +4.0	-0.8 to +2.3	0	-1.6 to +2.0	-1.9 to +3.0	-2.0 to +5.2	-1.6 to +8.1

Metal film resistors

Temperature coefficient: ±100 ppm/$^\circ$C

Wire-wound resistors

Temperature coefficient: $R \leq 1\Omega$ ±90 ppm/$^\circ$C
$1\Omega < R \leq 10\Omega$ ±50 ppm/$^\circ$C
$10\Omega < R$ ±20 ppm/$^\circ$C

Power Derating of Resistors with respect to Ambient Temperature

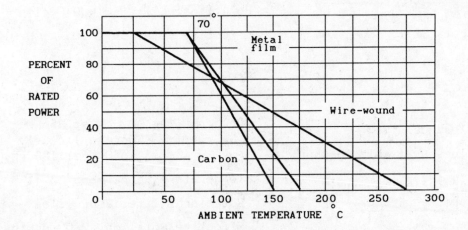

POTENTIOMETERS

CARBON COMPOSITION, Linear Taper, ± 10% Tolerance

Standard values in ohms

50	1 k	10k	100k	1 M
150	1.5k	15k	150k	1.5M
200	2 k	20k	200k	2 M
250	2.5k	25k	250k	2.5M
350	3.5k	35k	350k	3.5M
500	5 k	50k	500k	5 M
750	7.5k	75k	750k	

Power Rating: 2.25 W at 70 $^\circ$C
       Derate linearly
       to 0 at 120 $^\circ$C

Rotational Life: 100,000 cycles

Mechanical Rotation: 295° ±5°

Cost: $3.50

Temperature characteristics:
  Maximum % temporary resistance change from 25°C value.

Nominal	Temperature in $^\circ$C						
Resistance	-55	-25	0	25	55	85	120
R < 10 kΩ	+5.5	+3.0	+1.5	0	±1.0	±2.0	±4.5
10 kΩ ≤ R	+8.0	+4.0	+2.0	0	±1.5	±3.0	±6.0

CONDUCTIVE PLASTIC

Tolerance: ±10%

Values available in ohms

250			
1 k	10k	100k	1 M
2.5k	25k	250k	2.5M
5 k	50k	500k	5 M

Power Rating: 0.5 W at 70 $^\circ$C
       Derate linearly
       to 0 W at 120 $^\circ$C

Temperature Range: -55 to 120 $^\circ$C

Rotational Life: 50,000 Cycles

Effective Rotation: 265° +0° -10°

Cost: $2.75

CERMET

Tolerance: ±10%

Values available in ohms

50				
100	1k	10k	100k	1M
200	2k	20k	200k	2M
500	5k	50k	500k	

Power Rating: 0.5 W at 85 $^\circ$C
       Derate linearly
       to 0 W at 125 $^\circ$C

Temperature Range: -55 to +125 $^\circ$C

Rotational life: 200 cycles

Effective Rotation: 280° nominal

Temperature Coefficient: ±100 ppm/$^\circ$C

       referenced to 25 $^\circ$C

Cost: $0.79

CAPACITORS

POLARIZED ALUMINUM ELECTROLYTIC CAPACITORS

Voltage (V)	Capacitance (μF)	Unit Cost ($)
10	22	0.17
	33	.18
	47	.20
	100	.21
	220	.27
	330	.34
	470	.41
	1000	.59
	2200	.94
	3300	1.25
	4700	1.66
	6800	2.41
	10000	3.31
25	10	0.17
	22	.20
	33	.21
	47	.22
	100	.27
	220	.42
	330	.52
	470	.66
	1000	.97
	2200	1.58
	3300	2.32
	4700	3.05

Voltage (V)	Capacitance (μF)	Unit Cost ($)
50	0.1	0.17
	.22	.17
	.33	.17
	.47	.17
	1.0	.17
	2.2	.17
	3.3	.17
	4.7	.17
	10	.20
	22	.22
	33	.26
	47	.29
	100	.44
	220	.66
	330	.88
	470	1.08
	1000	1.66
	2200	3.14

Tolerance: -10% to +50%

Temperature Range: -55°C to $+105^\circ$C

CERAMIC DISC CAPACITORS

Voltage (V)	Capacitance (pF)	Unit Cost ($)
200	10	
	15	
	22	
	33	
	47	
	68	
	100	0.35
	150	
	220	
	330	
	470	
	680	
	1000	
	1500	
	2200	
	3300	
	4700	0.65
	6800	
	10000	
	15000	

Tolerance: ±10%

Temperature Range: -55°C to $+125^\circ$C

Temperature Characterisitc:
$\pm$ 30 ppm/$^\circ$C
referenced to 25°C

MYLAR POLYESTER CAPACITORS

Tolerance: ± 10%

Temperature Range: -55°C to $+125\overset{\circ}{\text{C}}$

Temperature Characteristic

	Temperature $^{\circ}$C				
	-55	25	80	100	125
Percent change in capacitance	-2	0	+1.5	+3	+9

Voltage (V)	Capacitance (μF)	Unit Cost ($)
100	0.001	0.46
	.0015	.46
	.0022	.47
	.0033	.47
	.0047	.47
	.0068	.47
	.0082	.47
	.01	.46
	.015	.62
	.022	.46
	.027	.46
	.033	.46
	.039	.51
	.047	.51
	.056	.54
	.068	.54
	.082	.54
	.1	.54
	.12	.72
	.15	.72
	.18	.85
	.22	.85
	.27	1.02
	.33	1.02
	.39	1.62
	.47	1.62
	.56	1.99
	.68	1.99
	.82	2.47
	1	2.47

CERAMIC VARIABLE CAPACITORS

Voltage (V)	Capacitance (pF) min	max	Unit Cost
250	1	4.5	2.95
	2.5	10	
	4	18	
	6	35	
	7	40	
	8	50	

INDUCTORS

Molded RF Chokes
±10 % Tolerance
Power Rating, 1/4 W

Inductance µH ±10%	Min Q	Test Freq MHz	Unit Cost
0.10	50	25	
.12	51	25	
.15	51	25	
.18	50	25	
.22	49	25	
.27	47	25	2.33
.33	46	25	
.39	44	25	
.47	44	25	
.56	43	25	
.68	42	25	
.82	40	25	
1.0	44	25	
1.2	44	7.9	
1.5	44	7.9	
1.8	44	7.9	
2.2	44	7.9	
2.7	44	7.9	
3.3	44	7.9	
3.9	44	7.9	
4.7	44	7.9	
5.6	44	7.9	2.44
6.8	50	7.9	
8.2	50	7.9	
10	50	7.9	
12	55	7.9	
15	45	2.5	
18	45	2.5	
22	45	2.5	
27	45	2.5	
33	45	2.5	
39	50	2.5	
47	50	2.5	2.68
56	50	2.5	
68	50	2.5	
82	50	2.5	
100	50	2.5	
120	55	.79	
150	55	.79	
180	55	.79	
220	55	.79	
270	55	.79	
330	55	.79	2.93
390	60	.79	
470	60	.79	
560	60	.79	
680	60	.79	
820	60	.79	
1000	60	.79	
1200	45	.25	
1500	45	.25	
1800	45	.25	
2200	45	.25	3.36
2700			
3300	45	.25	
3900	45	.25	
4700	45	.25	
5600	44	.25	
6800	40	.25	3.62
8200	40	.25	
10,000	40	.25	
15,000	30	.079	
18,000	30	.079	
22,000	27	.079	
27,000	27	.079	
33,000	27	.079	4.01
39,000	27	.079	
47,000	23	.079	
56,000	23	.079	
68,000	23	.079	4.75
100,000	18	.079	

Fixed Wirewound Inductors
±20 % Tolerance
0.1 A Maximum Current

Inductance H ±20%	Q	dc Res. (Ω)	Unit Cost
0.10	18	31.0	
.12	17	31.5	6.38
.15	17	32.0	
.22	16	33.0	
.27	16	33.5	
.33	16	34.0	
.39	15	35.0	
.47	15	35.5	7.00
.56	15	36.0	
.68	14	37.0	
.82	14	37.5	
1.0	14	38.0	
1.2	14	38.5	
1.5	13	39.0	
1.8	13	40.0	8.00
2.2	13	41.0	
2.7	13	42.0	
3.3	12	42.5	
3.9	12		
4.7	12	43.0	
5.6	12	44.0	9.00
6.8	10	44.5	
8.2	10	45.0	

Discrete Electronic Devices

DISCRETE ELECTRONIC DEVICES - COST DATA

SILICON DIODES

1N914	$ .55
1N914 A	$ .20
1N914 B	$ .20
1N916	$ .30
1N916 A	$ .30
1N916B	$ .30
1N9154	$ .30
1N9154	$ .30
1N4001, 1N4002	$ .15
1N4003, 1N4004	$ .20
1N4005, 1N4006, 1N4007	$ .25

ZENER DIODES

1N746 thru 1N759	$ .40
1N957A thru 1N973A	$ .40
1N974A thru 1N978A	$ .55
1N979A thru 1N983A	$1.10
1N983A thru 1N986A	$1.50
1N4370, 1N4371, 1N4372	$ .45

BIPOLAR JUNCTION TRANSISTORS

2N3903	$ .10
2N3904	$ .20
2N3905	$ .10
2N3906	$ .20
MJE171, MJE181	$ .95
MJE172, MJE182	$1.05

FIELD-EFFECT TRANSISTORS

2N5484	$ .30
2N5485	$ .30
2N5486	$ .30

OPTO-ELECTRONIC DEVICES

LED 55B, BF	$1.40
LED 55C, CF	$1.50
LED 56, 56F	$1.45
2N5777, 79	$1.75
2N5778, 80	$1.50
MLED 76	$1.45

Silicon
Diodes

⊳P ▮N	
1N914,A,B	
1N916,A,B	
1N4148,49	
1N4154	
1N4446-49	
1N4531	
1N4536	

This family of General Electric silicon signal diodes are very high speed switching diodes for computer circuits and general purpose applications. These diodes incorporate an oxide passivated planar structure. This structure makes possible a diode having high conductance, fast recovery time, low leakage, and low capacitance combined with improved uniformity and reliability. These diodes are contained in two different packages; double heat sink miniature package, and milli-heat sink package.

They are electrically the same as their equivalent types in each of the two different packages (see page two for groupings of electrically equivalent types in each of the two packages).

PLANAR EPITAXIAL PASSIVATED
with Controlled Conductance

absolute maximum ratings: (25°C) (unless otherwise specified)

		MHD & DHD	MHD & DHD	
Voltage	Reverse	75	25	Volts
Current	Average Rectified	150	150	mA
	Recurrent Peak Forward	450	450	mA
	Forward Steady-State DC	200	200	mA
	Peak Forward Surge (1μsec. pulse)	2000	2000	mA
Power	Dissipation	500	500	mW
Temperature	Operating	←−65 to +200→		°C
	Storage	←−65 to +200→		°C

electrical characteristics: (25°C) (unless otherwise specified)

Type	Minimum Breakdown Voltage @ 100μA	Forward Voltage		Maximum Reverse Current, I_R			$C_o^{(1)}$	$t_{rr}^{(2)}$	$V_f^{(3)}$
		I_F	V_F	20V		75V			
				25°C	150°C	25°C			
	Volts	mA	V	nA	μA	μA	pF	ns	V
1N914 1N4148 1N4531	100	10	1.0	25	50	5	4	4	
1N914A 1N4446	100	20	1.0	25	50	5	4	4	
1N914B 1N4448	100	} 5 } 100	0.62–0.72 { 1.0 {	25$^{(4)}$	50	5	4	4	2.5
1N916 1N4149	100	10	1.0	25	50	5	2	4	
1N916A 1N4447	100	20	1.0	25	50	5	2	4	
1N916B 1N4449	100	} 5 } 30	0.63–0.73 { 1.0 {	25	50	5	2	4	2.5
1N4154 1N4536	35 @ 5μA	30	1.0	100 @ 25V	100 @ 25V		4	2	

*Except as noted.

From G.E. Semiconductor Data Handbook, 3/e, © 1977. Reprinted with permission of Harris Semiconductor.

1N**4001** (SILICON)
thru
1N**4007**

Designers Data Sheet

"SURMETIC" RECTIFIERS

. . . subminiature size, axial lead mounted rectifiers for general-purpose low-power applications.

Designers Data for "Worst Case" Conditions

The Designers Data Sheets permit the design of most circuits entirely from the information presented. Limit curves — representing boundaries on device characteristics — are given to facilitate "worst case" design.

LEAD MOUNTED SILICON RECTIFIERS

50-1000 VOLTS DIFFUSED JUNCTION

*MAXIMUM RATINGS

Rating	Symbol	1N4001	1N4002	1N4003	1N4004	1N4005	1N4006	1N4007	Unit
Peak Repetitive Reverse Voltage Working Peak Reverse Voltage DC Blocking Voltage	V_{RRM} V_{RWM} V_R	50	100	200	400	600	800	1000	Volts
Non-Repetitive Peak Reverse Voltage (halfwave, single phase, 60 Hz)	V_{RSM}	60	120	240	480	720	1000	1200	Volts
RMS Reverse Voltage	$V_{R(RMS)}$	35	70	140	280	420	560	700	Volts
Average Rectified Forward Current (single phase, resistive load, 60 Hz, see Figure 8, $T_A = 75^oC$)	I_O				1.0				Amp
Non-Repetitive Peak Surge Current (surge applied at rated load conditions, see Figure 2)	I_{FSM}				30 (for 1 cycle)				Amp
Operating and Storage Junction Temperature Range	T_J, T_{stg}				-65 to $+175$				^oC

*ELECTRICAL CHARACTERISTICS

Characteristic and Conditions	Symbol	Typ	Max	Unit
Maximum Instantaneous Forward Voltage Drop ($i_F = 1.0$ Amp, $T_J = 25^oC$) Figure 1	v_F	0.93	1.1	Volts
Maximum Full-Cycle Average Forward Voltage Drop ($I_O = 1.0$ Amp, $T_L = 75^oC$, 1 inch leads)	$V_{F(AV)}$	—	0.8	Volts
Maximum Reverse Current (rated dc voltage) $T_J = 25^oC$ $T_J = 100^oC$	I_R	0.05 1.0	10 50	μA
Maximum Full-Cycle Average Reverse Current ($I_O = 1.0$ Amp, $T_L = 75^oC$, 1 inch leads	$I_{R(AV)}$	—	30	μA

*Indicates JEDEC Registered Data.

MECHANICAL CHARACTERISTICS

CASE: Void free, Transfer Molded
MAXIMUM LEAD TEMPERATURE FOR SOLDERING PURPOSES: 350^oC, 3/8" from case for 10 seconds at 5 lbs. tension
FINISH: All external surfaces are corrosion-resistant, leads are readily solderable
POLARITY: Cathode indicated by color band
WEIGHT: 0.40 Grams (approximately)

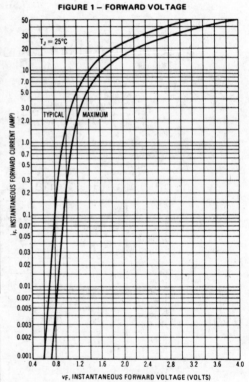

FIGURE 1 – FORWARD VOLTAGE

i_F, INSTANTANEOUS FORWARD CURRENT (AMP)

v_F, INSTANTANEOUS FORWARD VOLTAGE (VOLTS)

MOTOROLA
■ **SEMICONDUCTOR** ■
TECHNICAL DATA

Designers Data Sheet

500-MILLIWATT HERMETICALLY SEALED GLASS SILICON ZENER DIODES

- Complete Voltage Range — 2.4 to 110 Volts
- DO-35 Package — Smaller than Conventional DO-7 Package
- Double Slug Type Construction
- Metallurgically Bonded Construction
- Oxide Passivated Die

Designer's Data for "Worst Case" Conditions

The Designer's Data sheets permit the design of most circuits entirely from the information presented. Limit curves — representing boundaries on device characteristics — are given to facilitate "worst case" design.

1N746 thru 1N759
1N957A thru 1N986A
1N4370 thru 1N4372

GLASS ZENER DIODES
500 MILLIWATTS
2.4–110 VOLTS

MAXIMUM RATINGS

Rating	Symbol	Value	Unit
DC Power Dissipation @ $T_L \leq 50^\circ C$, Lead Length = 3/8"	P_D		
*JEDEC Registration		400	mW
*Derate above $T_L = 50^\circ C$.		3.2	mW/$^\circ$C
Motorola Device Ratings		500	mW
Derate above $T_L = 50^\circ C$		3.33	mW/$^\circ$C
Operating and Storage Junction Temperature Range	T_J, T_{stg}		$^\circ$C
*JEDEC Registration		−65 to +175	
Motorola Device Ratings		−65 to +200	

*Indicates JEDEC Registered Data.

MECHANICAL CHARACTERISTICS

MAXIMUM LEAD TEMPERATURE FOR SOLDERING PURPOSES: 230°C, 1/16" from case for 10 seconds

FINISH: All external surfaces are corrosion resistant with readily solderable leads.

POLARITY: Cathode indicated by color band. When operated in zener mode, cathode will be positive with respect to anode.

MOUNTING POSITION: Any

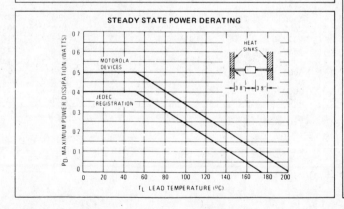

STEADY STATE POWER DERATING

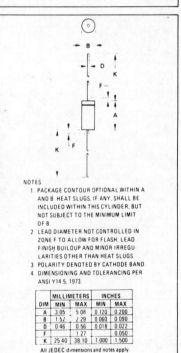

NOTES
1. PACKAGE CONTOUR OPTIONAL WITHIN A AND B. HEAT SLUGS, IF ANY, SHALL BE INCLUDED WITHIN THIS CYLINDER, BUT NOT SUBJECT TO THE MINIMUM LIMIT OF B.
2. LEAD DIAMETER NOT CONTROLLED IN ZONE F TO ALLOW FOR FLASH, LEAD FINISH BUILDUP AND MINOR IRREGULARITIES OTHER THAN HEAT SLUGS.
3. POLARITY DENOTED BY CATHODE BAND.
4. DIMENSIONING AND TOLERANCING PER ANSI Y14.5, 1973.

DIM	MILLIMETERS		INCHES	
	MIN	MAX	MIN	MAX
A	3.05	5.08	0.120	0.200
B	1.52	2.29	0.060	0.090
D	0.46	0.56	0.018	0.022
F		1.27		0.050
K	25.40	38.10	1.000	1.500

All JEDEC dimensions and notes apply.

CASE 299-02
DO-204AH
GLASS

Copyright of Motorola, Inc. Used by permission.

1N746 thru 1N759, 1N957A thru 1N986A, 1N4370 thru 1N4372

ELECTRICAL CHARACTERISTICS (T_A = 25°C, V_F = 1.5 V max at 200 mA for all types)

Type Number (Note 1)	Nominal Zener Voltage V_Z @ I_{ZT} (Note 2) Volts	Test Current I_{ZT} mA	Maximum Zener Impedance Z_{ZT} @ I_{ZT} (Note 3) Ohms	*Maximum DC Zener Current I_{ZM} (Note 4) mA		Maximum Reverse Leakage Current T_A = 25°C I_R @ V_R = 1 V µA	T_A = 150°C I_R @ V_R = 1 V µA
1N4370	2.4	20	30	150	190	100	200
1N4371	2.7	20	30	135	165	75	150
1N4372	3.0	20	29	120	150	50	100
1N746	3.3	20	28	110	135	10	30
1N747	3.6	20	24	100	125	10	30
1N748	3.9	20	23	95	115	10	30
1N749	4.3	20	22	85	105	2	30
1N750	4.7	20	19	75	95	2	30
1N751	5.1	20	17	70	85	1	20
1N752	5.6	20	11	65	80	1	20
1N753	6.2	20	7	60	70	0.1	20
1N754	6.8	20	5	55	65	0.1	20
1N755	7.5	20	6	50	60	0.1	20
1N756	8.2	20	8	45	55	0.1	20
1N757	9.1	20	10	40	50	0.1	20
1N758	10	20	17	35	45	0.1	20
1N759	12	20	30	30	35	0.1	20

Type Number (Note 1)	Nominal Zener Voltage V_Z (Note 2) Volts	Test Current I_{ZT} mA	Maximum Zener Impedance (Note 3) Z_{ZT} @ I_{ZT} Ohms	Z_{ZK} @ I_{ZK} Ohms	I_{ZK} mA	*Maximum DC Zener Current I_{ZM} (Note 4) mA		Maximum Reverse Current I_R Maximum µA	Test Voltage Vdc 5% V_R	10%
1N957A	6.8	18.5	4.5	700	1.0	47	61	150	5.2	4.9
1N958A	7.5	16.5	5.5	700	0.5	42	55	75	5.7	5.4
1N959A	8.2	15	6.5	700	0.5	38	50	50	6.2	5.9
1N960A	9.1	14	7.5	700	0.5	35	45	25	6.9	6.6
1N961A	10	12.5	8.5	700	0.25	32	41	10	7.6	7.2
1N962A	11	11.5	9.5	700	0.25	28	37	5	8.4	8.0
1N963A	12	10.5	11.5	700	0.25	26	34	5	9.1	8.6
1N964A	13	9.5	13	700	0.25	24	32	5	9.9	9.4
1N965A	15	8.5	16	700	0.25	21	27	5	11.4	10.8
1N966A	16	7.8	17	700	0.25	19	37	5	12.2	11.5
1N967A	18	7.0	21	750	0.25	17	23	5	13.7	13.0
1N968A	20	6.2	25	750	0.25	15	20	5	15.2	14.4
1N969A	22	5.6	29	750	0.25	14	18	5	16.7	15.8
1N970A	24	5.2	33	750	0.25	13	17	5	18.2	17.3
1N971A	27	4.6	41	750	0.25	11	15	5	20.6	19.4
1N972A	30	4.2	49	1000	0.25	10	13	5	22.8	21.6
1N973A	33	3.8	58	1000	0.25	9.2	12	5	25.1	23.8
1N974A	36	3.4	70	1000	0.25	8.5	11	5	27.4	25.9
1N975A	39	3.2	80	1000	0.25	7.8	10	5	29.7	28.1
1N976A	43	3.0	93	1500	0.25	7.0	9.6	5	32.7	31.0
1N977A	47	2.7	105	1500	0.25	6.4	8.8	5	35.8	33.8
1N978A	51	2.5	125	1500	0.25	5.9	8.1	5	38.8	36.7
1N979A	56	2.2	150	2000	0.25	5.4	7.4	5	42.6	40.3
1N980A	62	2.0	185	2000	0.25	4.9	6.7	5	47.1	44.6
1N981A	68	1.8	230	2000	0.25	4.5	6.1	5	51.7	49.0
1N982A	75	1.7	270	2000	0.25	1.0	5.5	5	56.0	54.0
1N983A	82	1.5	330	3000	0.25	3.7	5.0	5	62.2	59.0
1N984A	91	1.4	400	3000	0.25	3.3	4.5	5	69.2	65.5
1N985A	100	1.3	500	3000	0.25	3.0	4.5	5	76	72
1N986A	110	1.1	750	4000	0.25	2.7	4.1	5	83.6	79.2

NOTE 1. TOLERANCE AND VOLTAGE DESIGNATION

Tolerance Designation

The type numbers shown have tolerance designations as follows:

1N4370 series: ±10%, suffix A for ±5% units, C for ±2%, D for ±1%.

1N746 series: ±10%, suffix A for ±5% units, C for ±2%, D for ±1%.

1N957 series: ±10%, suffix A for ±10% units, C for ±2%, D for ±1%, suffix B for ±5% units, C for ±2%, D for ±1%.

Copyright of Motorola, Inc. Used by permission.

2N**5484** (SILICON)
thru
2N**5486**

N-channel depletion mode (Type A) junction field-effect transistors designed for VHF/UHF amplifier applications.

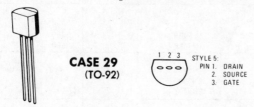

CASE 29
(TO-92)

STYLE 5:
PIN 1. DRAIN
2. SOURCE
3. GATE

MAXIMUM RATINGS

Rating	Symbol	Value	Unit
Drain-Gate Voltage	V_{DG}	25	Vdc
Reverse Gate-Source Voltage	$V_{GS(r)}$	25	Vdc
Drain Current	I_D	30	mAdc
Forward Gate Current	$I_{G(f)}$	10	mAdc
Total Device Dissipation @ $T_C = 25°C$ Derate above 25°C	$P_D^{(1)}$	310 2.82	mW mW/°C
Operating and Storage Junction Temperature Range	$T_J, T_{stg}^{(1)}$	-65 to +150	°C

(1) Continuous package improvements have enhanced these guaranteed Maximum Ratings as follows: $P_D = 1.0$ W @ $T_C = 25°C$. Derate above 25°C – 8.0 mW/°C, $T_J = -65$ to +150°C, $\theta_{JC} = 125°$ C/W.

ELECTRICAL CHARACTERISTICS ($T_C = 25°C$ unless otherwise noted)

Characteristic	Symbol	Min	Typ	Max	Unit
OFF CHARACTERISTICS					
Gate-Source Breakdown Voltage ($I_G = -1.0$ μAdc, $V_{DS} = 0$)	$V_{(BR)GSS}$	25	–	–	Vdc
Gate-Source Cutoff Voltage ($V_{DS} = 15$ Vdc, $I_D = 10$ nAdc) 2N5484 2N5485 2N5486	$V_{GS(off)}$	0.3 0.5 2.0	– – –	3.0 4.0 6.0	Vdc
Gate Reverse Current ($V_{GS} = -20$ Vdc, $V_{DS} = 0$)	I_{GSS}	–	–	1.0	nAdc
($V_{GS} = -20$ Vdc, $V_{DS} = 0$, $T_A = 100°C$)		–	–	0.2	μAdc
ON CHARACTERISTICS					
Zero-Gate Voltage Drain Current ($V_{DS} = 15$ Vdc, $V_{GS} = 0$) 2N5484 2N5485 2N5486	I_{DSS}	1.0 4.0 8.0	– – –	5.0 10 20	mAdc

2N5484 thru 2N5486 (continued)

ELECTRICAL CHARACTERISTICS (continued)

Characteristic	Symbol	Min	Typ	Max	Unit
SMALL-SIGNAL CHARACTERISTICS					
Forward Transadmittance (V_{DS} = 15 Vdc, V_{GS} = 0, f = 1.0 kHz) 2N5484 2N5485 2N5486	$|y_{fs}|$	3000 3500 4000	- - -	6000 7000 8000	μmhos
Forward Transconductance (V_{DS} = 15 Vdc, V_{GS} = 0, f = 100 MHz) 2N5484 (V_{DS} = 15 Vdc, V_{GS} = 0, f = 400 MHz) 2N5485 2N5486	$Re(y_{fs})$	2500 3000 3500	- - -	- - -	μmhos
Output Admittance (V_{DS} = 15 Vdc, V_{GS} = 0, f = 1.0 kHz) 2N5484 2N5485 2N5486	$|y_{os}|$	- - -	- - -	50 60 75	μmhos
Output Conductance (V_{DS} = 15 Vdc, V_{GS} = 0, f = 100 MHz) 2N5484 (V_{DS} = 15 Vdc, V_{GS} = 0, f = 400 MHz) 2N5485, 2N5486	$Re(y_{os})$	- -	- -	75 100	μmhos
Input Conductance (V_{DS} = 15 Vdc, V_{GS} = 0, f = 100 MHz) 2N5484 (V_{DS} = 15 Vdc, V_{GS} = 0, f = 400 MHz) 2N5485, 2N5486	$Re(y_{is})$	- -	- -	100 1000	μmhos
Input Capacitance (V_{DS} = 15 Vdc, V_{GS} = 0, f = 1.0 MHz)	C_{iss}	-	-	5.0	pF
Reverse Transfer Capacitance (V_{DS} = 15 Vdc, V_{GS} = 0, f = 1.0 MHz)	C_{rss}	-	-	1.0	pF
Output Capacitance (V_{DS} = 15 Vdc, V_{GS} = 0, f = 1.0 MHz)	C_{oss}	-	-	2.0	pF
Common-Source Noise Figure (V_{DS} = 15 Vdc, V_{GS} = 0, R_G = 1.0 Megohm, f = 1.0 kHz) All Types (V_{DS} = 15 Vdc, I_D = 1.0 mAdc, $R_G \approx$ 1.0 k ohm, f = 100 MHz) 2N5484 (V_{DS} = 15 Vdc, I_D = 1.0 mAdc, $R_G \approx$ 1.0 k ohm, f = 200 MHz) 2N5484 (V_{DS} = 15 Vdc, I_D = 4.0 mAdc, $R_G \approx$ 1.0 k ohm, f = 100 MHz) 2N5485, 2N5486 (V_{DS} = 15 Vdc, I_D = 4.0 mAdc, $R_G \approx$ 1.0 k ohm, f = 400 MHz) 2N5485, 2N5486	NF	- - - - -	- - 4.0 - -	2.5 3.0 - 2.0 4.0	dB
Insertion Power Gain (V_{DS} = 15 Vdc, I_D = 1.0 mAdc, f = 100 MHz) 2N5484 (V_{DS} = 15 Vdc, I_D = 1.0 mAdc, f = 200 MHz) 2N5484 (V_{DS} = 15 Vdc, I_D = 4.0 mAdc, f = 100 MHz) 2N5485, 2N5486 (V_{DS} = 15 Vdc, I_D = 4.0 mAdc, f = 400 MHz) 2N5485, 2N5486	G_{ps}	16 - 18 10	- 14 - -	25 - 30 20	dB

MAXIMUM RATINGS

Rating	Symbol	Value	Unit
Collector-Emitter Voltage	V_{CEO}	40	Vdc
Collector-Base Voltge	V_{CBO}	60	Vdc
Emitter-Base Voltage	V_{EBO}	6.0	Vdc
Collector Current — Continuous	I_C	200	mAdc
Total Device Dissipation @ T_A = 25°C Derate above 25°C	P_D	625 5.0	mW mW/°C
*Total Device Dissipation @ T_C = 25°C Derate above 25°C	P_D	1.5 12	Watts mW/°C
Operating and Storage Junction Temperature Range	T_J, T_{stg}	−55 to +150	°C

*THERMAL CHARACTERISTICS

Characteristic	Symbol	Max	Unit
Thermal Resistance, Junction to Case	$R_{\theta JC}$	83.3	°C/W
Thermal Resistance, Junction to Ambient	$R_{\theta JA}$	200	°C/W

*Indicates Data in addition to JEDEC Requirements.

2N3903
2N3904

CASE 29-04, STYLE 1
TO-92 (TO-226AA)

3 Collector

2 Base

1 Emitter

GENERAL PURPOSE
TRANSISTORS

NPN SILICON

ELECTRICAL CHARACTERISTICS (T_A = 25°C unless otherwise noted.)

Characteristic		Symbol	Min	Max	Unit
OFF CHARACTERISTICS					
Collector-Emitter Breakdown Voltage(1) (I_C = 1.0 mAdc, I_B = 0)		$V_{(BR)CEO}$	40	—	Vdc
Collector-Base Breakdown Voltage (I_C = 10 μAdc, I_E = 0)		$V_{(BR)CBO}$	60	—	Vdc
Emitter-Base Breakdown Voltage (I_E = 10 μAdc, I_C = 0)		$V_{(BR)EBO}$	6.0	—	Vdc
Base Cutoff Current (V_{CE} = 30 Vdc, V_{EB} = 3.0 Vdc)		I_{BL}	—	50	nAdc
Collector Cutoff Current (V_{CE} = 30 Vdc, V_{EB} = 3.0 Vdc)		I_{CEX}	—	50	nAdc
ON CHARACTERISTICS					
DC Current Gain(1) (I_C = 0.1 mAdc, V_{CE} = 1.0 Vdc)	2N3903 2N3904	h_{FE}	20 40	— —	—
(I_C = 1.0 mAdc, V_{CE} = 1.0 Vdc)	2N3903 2N3904		35 70	— —	
(I_C = 10 mAdc, V_{CE} = 1.0 Vdc)	2N3903 2N3904		50 100	150 300	
(I_C = 50 mAdc, V_{CE} = 1.0 Vdc)	2N3903 2N3904		30 60	— —	
(I_C = 100 mAdc, V_{CE} = 1.0 Vdc)	2N3903 2N3904		15 30	— —	

Collector-Emitter Saturation Voltage(1)
  (I_C = 10 mAdc, I_B = 1.0 mAdc)
  (I_C = 50 mAdc, I_B = 5.0 mAdc)
		$V_{CE(sat)}$	—
			—
				0.2
				0.3
					Vdc

Base-Emitter Saturation Voltage(1)
  (I_C = 10 mAdc, I_B = 1.0 mAdc)
  (I_C = 50 mAdc, I_B = 5.0 mAdc)
		$V_{BE(sat)}$	0.65
			—
				0.85
				0.95
					Vdc

SMALL-SIGNAL CHARACTERISTICS					
Current-Gain — Bandwidth Product (I_C = 10 mAdc, V_{CE} = 20 Vdc, f = 100 MHz)	2N3903 2N3904	f_T	250 300	— —	MHz

MOTOROLA SMALL-SIGNAL TRANSISTORS, FETs AND DIODES

2N3903, 2N3904

ELECTRICAL CHARACTERISTICS (continued) (T_A = 25°C unless otherwise noted.)

Characteristic		Symbol	Min	Max	Unit
Output Capacitance (V_{CB} = 5.0 Vdc, I_E = 0, f = 1.0 MHz)		C_{obo}	—	4.0	pF
Input Capacitance (V_{BE} = 0.5 Vdc, I_C = 0, f = 1.0 MHz)		C_{ibo}	—	8.0	pF
Input Impedance (I_C = 1.0 mAdc, V_{CE} = 10 Vdc, f = 1.0 kHz)	2N3903 2N3904	h_{ie}	1.0 1.0	8.0 10	k ohms
Voltage Feedback Ratio (I_C = 1.0 mAdc, V_{CE} = 10 Vdc, f = 1.0 kHz)	2N3903 2N3904	h_{re}	0.1 0.5	5.0 8.0	X 10^{-4}
Small-Signal Current Gain (I_C = 1.0 mAdc, V_{CE} = 10 Vdc, f = 1.0 kHz)	2N3903 2N3904	h_{fe}	50 100	200 400	—
Output Admittance (I_C = 1.0 mAdc, V_{CE} = 10 Vdc, f = 1.0 kHz)		h_{oe}	1.0	40	μmhos
Noise Figure (I_C = 100 μAdc, V_{CE} = 5.0 Vdc, R_S = 1.0 k ohms, f = 1.0 kHz)	2N3903 2N3904	NF	— —	6.0 5.0	dB

SWITCHING CHARACTERISTICS

			Symbol	Min	Max	Unit
Delay Time	(V_{CC} = 3.0 Vdc, V_{BE} = 0.5 Vdc, I_C = 10 mAdc, I_{B1} = 1.0 mAdc)		t_d	—	35	ns
Rise Time			t_r	—	35	ns
Storage Time	(V_{CC} = 3.0 Vdc, I_C = 10 mAdc, I_{B1} = I_{B2} = 1.0 mAdc)	2N3903 2N3904	t_s	— —	175 200	ns
Fall Time			t_f	—	50	ns

(1) Pulse Test: Pulse Width ≤ 300 μs, Duty Cycle ≤ 2.0%.

MAXIMUM RATINGS

Rating	Symbol	Value	Unit
Collector-Emitter Voltage	V_{CEO}	−40	Vdc
Collector-Base Voltage	V_{CBO}	−40	Vdc
Emitter-Base Voltage	V_{EBO}	−5.0	Vdc
Collector Current — Continuous	I_C	−200	mAdc
Total Device Dissipation @ T_A = 25°C Derate above 25°C	P_D	625 5.0	mW mW/°C
Total Power Dissipation @ T_A = 60°C	P_D	250	mW
Total Divice Dissipation @ T_C = 25°C Derate above 25°C	P_D	1.5 12	Watts mW/°C
Operating and Storage Junction Temperature Range	T_J, T_{stg}	−55 to +150	°C

*THERMAL CHARACTERISTICS

Characteristic	Symbol	Max	Unit
Thermal Resistance, Junction to Ambient	$R_{\theta JA}$	200	°C/W
Thermal Resistance, Junction to Case	$R_{\theta JC}$	83.3	°C/W

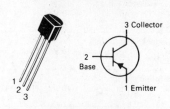

2N3905
2N3906★

**CASE 29-04, STYLE 1
TO-92 (TO-226AA)**

3 Collector

2 Base

1 Emitter

GENERAL PURPOSE
TRANSISTORS

PNP SILICON

★This is a Motorola
designated preferred device.

ELECTRICAL CHARACTERISTICS (T_A = 25°C unless otherwise noted.)

Characteristic		Symbol	Min	Max	Unit
OFF CHARACTERISTICS					
Collector-Emitter Breakdown Voltage (1) (I_C = −1.0 mAdc, I_B = 0)		$V_{(BR)CEO}$	−40	—	Vdc
Collector-Base Breakdown Voltage (I_C = −10 μAdc, I_E = 0)		$V_{(BR)CBO}$	−40	—	Vdc
Emitter-Base Breakdown Voltage (I_E = −10 μAdc, I_C = 0)		$V_{(BR)EBO}$	−5.0	—	Vdc
Base Cutoff Current (V_{CE} = −30 Vdc, V_{EB} = −3.0 Vdc)		I_{BL}	—	−50	nAdc
Collector Cutoff Current (V_{CE} = −30 Vdc, V_{EB} = −3.0 Vdc)		I_{CEX}	—	−50	nAdc
ON CHARACTERISTICS(1)					
DC Current Gain I_C = −0.1 mAdc, V_{CE} = −1.0 Vdc)	2N3905 2N3906	h_{FE}	 30 60	— — —	—
(I_C = −1.0 mAdc, V_{CE} = −1.0 Vdc)	2N3905 2N3906		40 80	— —	
(I_C = −10 mAdc, V_{CE} = −1.0 Vdc)	2N3905 2N3906		50 100	150 300	
(I_C = −50 mAdc, V_{CE} = −1.0 Vdc)	2N3905 2N3506		30 60	— —	
(I_C = −100 mAdc, V_{CE} = −1.0 Vdc)	2N3905 2N3906		15 30	— —	
Collector-Emitter Saturation Voltage (I_C = −10 mAdc, I_B = −1.0 mAdc) (I_C = −50 mAdc, I_B = −5.0 mAdc)		$V_{CE(sat)}$	 — —	 −0.25 −0.4	Vdc
Base-Emitter Saturation Voltage (I_C = −10 mAdc, I_B = −1.0 mAdc) (I_C = −50 mAdc, I_B = −5.0 mAdc)		$V_{BE(sat)}$	 −0.65 —	 −0.85 −0.95	Vdc
SMALL-SIGNAL CHARACTERISTICS					
Current-Gain — Bandwidth Product (I_C = −10 mAdc, V_{CE} = −20 Vdc, f = 100 MHz)	2N3905 2N3906	f_T	200 250	— —	MHz
Output Capacitance (V_{CB} = −5.0 Vdc, I_E = 0, f = 1.0 MHz)		C_{obo}	—	4.5	pF

MOTOROLA SMALL-SIGNAL TRANSISTORS, FETs AND DIODES

2N3905, 2N3906

ELECTRICAL CHARACTERISTICS (continued) (T_A = 25°C unless otherwise noted.)

Characteristic		Symbol	Min	Max	Unit
Input Capacitance (V_{EB} = −0.5 Vdc, I_C = 0, f = 1.0 MHz)		C_{ibo}	—	10.0	pF
Input Impedance (I_C = −1.0 mAdc, V_{CE} = −10 Vdc, f = 1.0 kHz)	2N3905 2N3906	h_{ie}	0.5 2.0	8.0 12	k ohms
Voltage Feedback Ratio (I_C = −1.0 mAdc, V_{CE} = −10 Vdc, f = 1.0 kHz	2N3905 2N3906	h_{re}	0.1 0.1	5.0 10	X 10^{-4}
Small-Signal Current Gain (I_C = −1.0 mAdc, V_{CE} = −10 Vdc, f = 1.0 kHz)	2N3905 2N3906	h_{fe}	50 100	200 400	—
Output Admittance (I_C = −1.0 mAdc, V_{CE} = −10 Vdc, f = 1.0 kHz)	2N3905 2N3906	h_{oe}	1.0 3.0	40 60	μmhos
Noise Figure (I_C = −100 μAdc, V_{CE} = −5.0 Vdc, R_S = 1.0 k ohm, f = 1.0 kHz)	2N3905 2N3906	NF	— —	5.0 4.0	dB

SWITCHING CHARACTERISTICS

			Symbol	Min	Max	Unit
Delay Time	(V_{CC} = −3.0 Vdc, V_{BE} = −0.5 Vdc		t_d	—	35	ns
Rise Time	I_C = −10 mAdc, I_{B1} = −1.0 mAdc)		t_r	—	35	ns
Storage Time		2N3905 2N3906	t_s	— —	200 225	ns
Fall Time	(V_{CC} = −3.0 Vdc, I_C = −10 mAdc, I_{B1} = I_{B2} = −1.0 mAdc)	2N3905 2N3906	t_f	— —	60 75	ns

(1) Pulse Width ≤ 300 μs, Duty Cycle ≤ 2.0%.

MOTOROLA
■ **SEMICONDUCTOR** ■
TECHNICAL DATA

PNP
MJE171*, MJE172*
NPN
MJE181*, MJE182*

*Motorola preferred devices

COMPLEMENTARY PLASTIC SILICON POWER TRANSISTORS

. . . designed for low power audio amplifier and low current, high speed switching applications.

- Collector-Emitter Sustaining Voltage —
  $V_{CEO(sus)}$ = 60 Vdc — MJE171, MJE181
  = 80 Vdc — MJE172, MJE182
- DC Current Gain —
  h_{FE} = 30 (Min) @ I_C = 0.5 Adc
  = 12 (Min) @ I_C = 1.5 Adc
- Current-Gain — Bandwidth Product —
  f_T = 50 MHz (Min) @ I_C = 100 mAdc
- Annular Construction for Low Leakages —
  I_{CBO} = 100 nA (Max) @ Rated V_{CB}

3 AMPERE
POWER TRANSISTORS
COMPLEMENTARY SILICON
60-80 VOLTS
12.5 WATTS

MAXIMUM RATINGS

Rating	Symbol	MJE171 MJE181	MJE172 MJE182	Unit
Collector-Base Voltage	V_{CB}	80	100	Vdc
Collector-Emitter Voltage	V_{CEO}	60	80	Vdc
Emitter-Base Voltage	V_{EB}	7.0		Vdc
Collector Current — Continuous Peak	I_C	3.0 6.0		Adc
Base Current	I_B	1.0		Adc
Total Power Dissipation @ T_A = 25°C Derate above 25°C	P_D	1.5 0.012		Watts W/°C
Total Power Dissipation @ T_C = 25°C Derate above 25°C	P_D	12.5 0.1		Watts W/°C
Operating and Storage Junction Temperature Range	T_J, T_{stg}	− 65 to + 150		°C

THERMAL CHARACTERISTICS

Characteristic	Symbol	Max	Unit
Thermal Resistance, Junction to Case	θ_{JC}	10	°C/W
Thermal Resistance, Junction to Ambient	θ_{JA}	83.4	°C/W

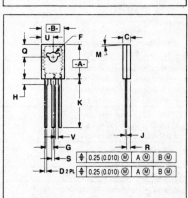

NOTES:
1. DIMENSIONING AND TOLERANCING PER ANSI Y14.5M, 1982.
2. CONTROLLING DIMENSION: INCH.
3. 077-01 THRU -06 OBSOLETE. NEW STANDARD 077-07.

DIM	MILLIMETERS		INCHES	
	MIN	MAX	MIN	MAX
A	10.80	11.04	0.425	0.435
B	7.50	7.74	0.295	0.305
C	2.42	2.66	0.095	0.105
D	0.51	0.66	0.020	0.026
F	2.93	3.30	0.115	0.130
G	2.39 BSC		0.094 BSC	
H	1.27	2.41	0.050	0.095
J	0.39	0.63	0.015	0.025
K	14.61	16.63	0.575	0.655
M	3 TYP		3 TYP	
Q	3.76	4.01	0.148	0.158
R	1.15	1.39	0.045	0.055
S	0.64	0.88	0.025	0.035
U	3.69	3.93	0.145	0.155
V	1.02	—	0.040	—

STYLE 1:
PIN 1. EMITTER
2. COLLECTOR
3. BASE

CASE 77-07
TO-225AA

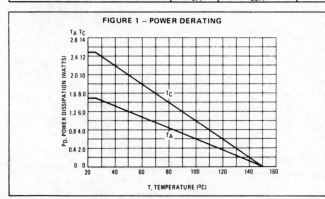

FIGURE 1 — POWER DERATING

P_D, POWER DISSIPATION (WATTS)

T_C

T_A

T, TEMPERATURE (°C)

Copyright of Motorola, Inc. Used by permission.

MJE171, MJE172, PNP, MJE181, MJE182, NPN

ELECTRICAL CHARACTERISTICS (T_C = 25°C unless otherwise noted)

Characteristic		Symbol	Min	Max	Unit
OFF CHARACTERISTICS					
Collector-Emitter Sustaining Voltage		$V_{CEO(sus)}$			Vdc
(I_C = 10 mAdc, I_B = 0) MJE171, MJE181			60	—	
MJE172, MJE182			80	—	
Collector Cutoff Current		I_{CBO}			μAdc
(V_{CB} = 80 Vdc, I_E = 0) MJE171, MJE181			—	0.1	
(V_{CB} = 100 Vdc, I_E = 0) MJE172, MJE182			—	0.1	
(V_{CB} = 80 Vdc, I_E = 0, T_C = 150°C) MJE171, MJE181			—	0.1	mAdc
(V_{CB} = 100 Vdc, I_E = 0, T_C = 150°C) MJE172, MJE182			—	0.1	
Emitter Cutoff Current		I_{EBO}	—	0.1	μAdc
(V_{BE} = 7.0 Vdc, I_C = 0)					
ON CHARACTERISTICS					
DC Current Gain		h_{FE}			—
(I_C = 100 mAdc, V_{CE} = 1.0 Vdc)			50	250	
(I_C = 500 mAdc, V_{CE} = 1.0 Vdc)			30	—	
(I_C = 1.5 Adc, V_{CE} = 1.0 Vdc)			12	—	
Collector-Emitter Saturation Voltage		$V_{CE(sat)}$			Vdc
(I_C = 500 mAdc, I_B = 50 mAdc)			—	0.3	
(I_C = 1.5 Adc, I_B = 150 mAdc)			—	0.9	
(I_C = 3.0 Adc, I_B = 600 mAdc)			—	1.7	
Base-Emitter Saturation Voltage		$V_{BE(sat)}$			Vdc
(I_C = 1.5 Adc, I_B = 150 mAdc)			—	1.5	
(I_C = 3.0 Adc, I_B = 600 mAdc)			—	2.0	
Base-Emitter On Voltage		$V_{BE(on)}$	—	1.2	Vdc
(I_C = 500 mAdc, V_{CE} = 1.0 Vdc)					
DYNAMIC CHARACTERISTICS					
Current-Gain — Bandwidth Product (1)		f_T	50	—	MHz
(I_C = 100 mAdc, V_{CE} = 10 Vdc, f_{test} = 10 MHz)					
Output Capacitance		C_{ob}			pF
(V_{CB} = 10 Vdc, I_E = 0, f = 0.1 MHz) MJE171/MJE172			—	60	
MJE181/MJE182			—	40	

(1) $f_T = |h_{fe}| \cdot f_{test}$

Copyright of Motorola, Inc. Used by permission.

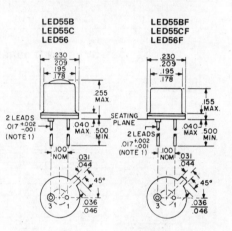

SOLID STATE **GENERAL ⊗ ELECTRIC**

OPTOELECTRONICS

Direct replacement for SSL55B, SSL55C, SSL56, SSL55BF, SSL55CF, SSL56F

Infrared Emitter

P N LED55B, LED55C, LED56, LED55BF, LED55CF, LED56F

Gallium Arsenide Infrared–Emitting Diode

The General Electric LED55B-LED55C-LED56 Series are gallium arsenide, light emitting diodes which emit non-coherent, infrared energy with a peak wave length of 940 nanometers. They are ideally suited for use with silicon detectors. The "F" versions of these devices have flat lens caps.

absolute maximum ratings: (25°C unless otherwise specified)

Voltage:
Reverse Voltage	V_R	3	volts

Currents:
Forward Current Continuous	I_F	100	mA
Forward Current (pw 1 μsec 200 Hz)	I_F	10	A

Dissipations:
Power Dissipation ($T_A = 25°C$)*	P_T	170	mW
Power Dissipation ($T_C = 25°C$)**	P_T	1.3	W

Temperatures:
Junction Temperature	T_J	-65°C to +150°C
Storage Temperature	T_{STG}	-65°C to +150°C
Lead Soldering Time		10 seconds at 260°C

*Derate 1.36 mW/°C above 25°C ambient.
**Derate 10.4 mW/°C above 25°C case.

electrical characteristics: (25°C unless otherwise specified)

		MIN.	TYP.	MAX.	UNITS
Reverse Leakage Current ($V_R = 3V$)	I_R			10	μA
Forward Voltage ($I_F = 100mA$)	V_F		1.4	1.7	V

optical characteristics: (25°C unless otherwise specified)

		MIN.	TYP.	MAX.	UNITS
Total Power Output (note 1) ($I_F = 100mA$)					
LED55B-LED55BF	P_O	3.5			mW
LED55C-LED55CF		5.4			mW
LED56 -LED56F		1.5			mW
Peak Emission Wavelength ($I_F = 100mA$)			940		nm
Spectral Shift with Temperature			.28		nm/°C
Spectral Bandwidth 50%			60		nm
Rise Time 0-90% of Output			300		nsec
Fall Time 100-10% of Output			200		nsec

Note 1: Total power output, P_O, is the total power radiated by the device into a solid angle of 2π steradians.

From G.E. Semiconductor Data Handbook, 3/e, © 1977. Reprinted with permission of Harris Semiconductor.

LED55B, LED55C, LED56, LED55BF, LED55CF, LED56F TYPICAL CHARACTERISTICS

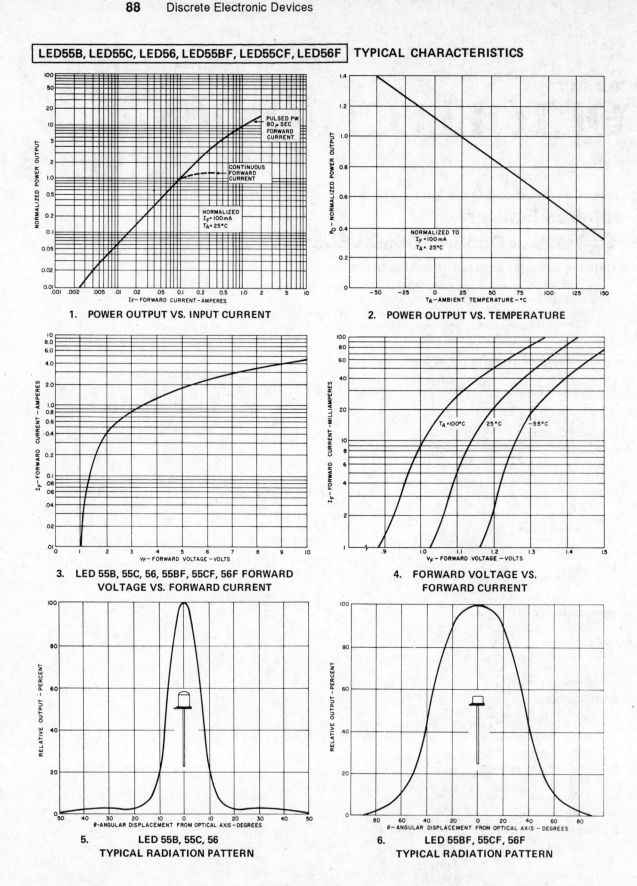

1. POWER OUTPUT VS. INPUT CURRENT

2. POWER OUTPUT VS. TEMPERATURE

3. LED 55B, 55C, 56, 55BF, 55CF, 56F FORWARD VOLTAGE VS. FORWARD CURRENT

4. FORWARD VOLTAGE VS. FORWARD CURRENT

5. LED 55B, 55C, 56 TYPICAL RADIATION PATTERN

6. LED 55BF, 55CF, 56F TYPICAL RADIATION PATTERN

SOLID STATE

GENERAL ELECTRIC

OPTOELECTRONICS

Light Detector Planar Silicon Photo-Darlington Amplifier

NPN 2N5777-80

This General Electric Light Sensor Series is an NPN planar silicon photo-darlington amplifier. For many applications, only the collector and emitter leads are used. A base lead is provided to control sensitivity and the gain of the device. They are packaged in clear epoxy encapsulant and can be used in industrial and commercial applications requiring a low-cost, general purpose, photosensitive device.

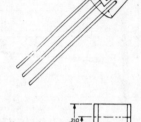

absolute maximum ratings: (25°C) (unless otherwise specified)

		2N5777, 79 (L14D1,3)	2N5778, 80 (L14D2,4)	
Voltages—Dark Characteristics				
Collector to Emitter	V_{CEO}	25	40	Volts
Collector to Base	V_{CBO}	25	40	Volts
Emitter to Base	V_{EBO}	8	12	Volts
Current				
Light Current	I_L	250	250	mA
Dissipation				
Power Dissipation*	P_T	200	200	mW
Temperature				
Junction Temperature	T_J	◄——— 100°C ———►		
Storage Temperature	T_{stg}	◄— −65°C to +100°C —►		

*Derate 2.67mW/°C above 25°C ambient

electrical characteristics: (25°C) (unless otherwise specified)

		2N5777, 78		2N5779, 80		
Static Characteristics		Min.	Max.	Min.	Max.	
Light Current (V_{CE} = 5V, H = 2mW/cm²**)	I_L	0.5	–	2.0	–	mA
Forward Current Transfer Ratio (V_{CE} = 5V, I_C = 2.0mA)	h_{FE}	1.0k	–	2.0k	–	

		2N5777, 79		2N5778, 80		
		Min.	Max.	Min.	Max.	
Dark Current (V_{CE} = 12V, I_B = 0)	I_D	–	100	–	100	nA
Collector-Emitter Breakdown Voltage (I_C = 10mA, H = 0)	$V_{(BR)CEO}$	25	–	40	–	Volts
Collector-Base Breakdown Voltage (I_C = 100μA, H = 0)	$V_{(BR)CBO}$	25	–	40	–	Volts
Emitter-Base Breakdown Voltage (I_E = 100μA, H = 0)	$V_{(BR)EBO}$	8	–	12	–	Volts

		2N5777-80			
Dynamic Characteristics		Min.	Typ.	Max.	
Switching Speeds (V_{CE} = 10V, I_L = 10mA, R_L = 100 ohms, GaAs LED source)					
Delay Time	t_d	–	30	100	μsec.
Rise Time	t_r	–	75	250	μsec.
Storage Time	t_s	–	0.5	5	μsec.
Fall Time	t_f	–	45	150	μsec.
Collector-Base Capacitance (V_{CB} = 10V, f = 1MHz)	C_{cb}	–	7.6	10	pF
Emitter-Base Capacitance (V_{EB} = 0.5V, f = 1MHz)	C_{eb}	–	10.5	–	pF
Collector-Emitter Capacitance (V_{CEO} = 10V, f - 1MHz)	C_{ceo}	–	3.4	–	pF

**H = Radiation Flux Density. Radiation source is an unfiltered tungsten filament bulb at 2870°K color temperature.

508

PELLET LOCATION

From G.E. Semiconductor Data Handbook, 3/e, © 1977. Reprinted with permission of Harris Semiconductor.

TYPICAL ELECTRICAL CHARACTERISTICS

2N5777,80

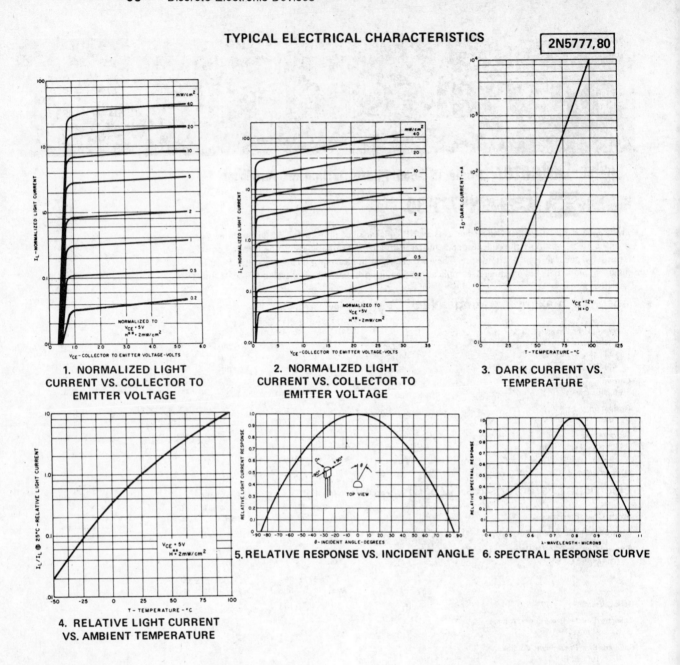

1. NORMALIZED LIGHT CURRENT VS. COLLECTOR TO EMITTER VOLTAGE

2. NORMALIZED LIGHT CURRENT VS. COLLECTOR TO EMITTER VOLTAGE

3. DARK CURRENT VS. TEMPERATURE

4. RELATIVE LIGHT CURRENT VS. AMBIENT TEMPERATURE

5. RELATIVE RESPONSE VS. INCIDENT ANGLE

6. SPECTRAL RESPONSE CURVE

Order this data sheet by MLED76

MOTOROLA
SEMICONDUCTOR
TECHNICAL DATA

Visible Red LED

This device is designed for a wide variety of applications where visible light emission is desirable, and can be used in conjunction with any MRD700 series detector. The MLED76 features high power output, using gallium aluminum arsenide technology.

- Low Cost
- Popular Case 349 Package
- Uses Stable Long-Life LED Technology
- Clear Epoxy Package

MLED76

VISIBLE RED LED 660 nm

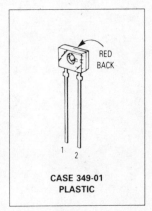

RED BACK

CASE 349-01 PLASTIC

MAXIMUM RATINGS

Rating	Symbol	Value	Unit
Reverse Voltage	V_R	5	Volts
Forward Current — Continuous	I_F	60	mA
Forward Current — Peak Pulse	I_F	1	A
Total Power Dissipation @ T_A = 25°C (Note 1) Derate above 35°C	P_D	132 2	mW mW/°C
Ambient Operating Temperature Range	T_A	40 to +100	°C
Storage Temperature	T_{stg}	40 to +100	°C
Lead Soldering Temperature (Note 2)	—	260	°C

ELECTRICAL CHARACTERISTICS (T_A = 25°C unless otherwise noted)

Characteristic	Symbol	Min	Typ	Max	Unit
Reverse Leakage Current (V_R = 3 V)	I_R	—	100	—	nA
Reverse Leakage Current (V_R = 5 V)	I_R	—	10	100	μA
Forward Voltage (I_F = 60 mA)	V_F	—	1.8	2.2	V
Temperature Coefficient of Forward Voltage	ΔV_F	—	-2.2	—	mV/K
Capacitance (f = 1 MHz)	C	—	50	—	pF

OPTICAL CHARACTERISTICS (T_A = 25°C unless otherwise noted)

Characteristic	Symbol	Min	Typ	Max	Unit
Peak Wavelength (I_F = 60 mA)	λp	—	660	—	nm
Spectral Half-Power Bandwidth	$\Delta\lambda$	—	20	—	nm
Continuous Power Output (I_F = 60 mA) (Note 3)	P_O	—	2.2	—	mW
Instantaneous Power Output (I_F = 100 mA)	P_O	—	4	—	mW
Instantaneous Axial Intensity (I_F = 100 mA) (Note 4)	I_O	0.8	1.3	—	mW/sr
Power Half-Angle	θ	—	±30	—	
Optical Turn-On Time	t_{on}	—	200	—	ns
Optical Turn-Off Time	t_{off}	—	150	—	ns
Half-Power Electrical Bandwidth (Note 5)	BWe	—	6	—	MHz

Notes: 1. Measured with device soldered into a typical printed circuit board.
    2. 5 seconds max; 1/16 inch from case. Heat sink should be applied during soldering, to prevent case temperature from exceeding 100°C.
    3. Measured using a Photodyne 88xLA with a #350 integrating sphere.
    4. On-axis, with cone angle of ·13°.
    5. I_F = 100 mA pk-pk, 100% modulation.

MOTOROLA

DS2741

Linear Integrated Circuits

LINEAR INTEGRATED CIRCUITS - COST DATA

PART #	PRICE
Operational Amplifiers (Uncompensated)	
LM108	$5.14
LM208	3.86
LM308	.72
Voltage Comparators	
LM111	$4.45
LM211	3.68
LM311	.42
Adjustable Voltage Regulators	
LM117	$15.05
LM317	2.69
LM317T	.80
LM317M	NA
Timers	
LM555	$4.80
LM555C	1.30
Operational Amplifiers (Compensated)	
LM741	$1.54
LM741A	1.89
LM741C	.54
LM741E	.89

 **National Semiconductor**

May 1989

LM108/LM208/LM308 Operational Amplifiers

General Description

The LM108 series are precision operational amplifiers having specifications a factor of ten better than FET amplifiers over a $-55°C$ to $+125°C$ temperature range.

The devices operate with supply voltages from $\pm2V$ to $\pm20V$ and have sufficient supply rejection to use unregulated supplies. Although the circuit is interchangeable with and uses the same compensation as the LM101A, an alternate compensation scheme can be used to make it particularly insensitive to power supply noise and to make supply bypass capacitors unnecessary.

The low current error of the LM108 series makes possible many designs that are not practical with conventional amplifiers. In fact, it operates from 10 MΩ source resistances,

introducing less error than devices like the 709 with 10 kΩ sources. Integrators with drifts less than 500 μV/sec and analog time delays in excess of one hour can be made using capacitors no larger than 1 μF.

The LM108 is guaranteed from $-55°C$ to $+125°C$, the LM208 from $-25°C$ to $+85°C$, and the LM308 from $0°C$ to $+70°C$.

Features

- Maximum input bias current of 3.0 nA over temperature
- Offset current less than 400 pA over temperature
- Supply current of only 300 μA, even in saturation
- Guaranteed drift characteristics

Compensation Circuits

Standard Compensation Circuit

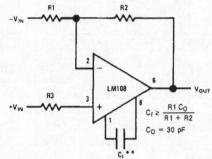

$$C_f \geq \frac{R1\,C_O}{R1 + R2}$$

$$C_O = 30\ pF$$

TL/H/7758–1

**Bandwidth and slew rate are proportional to $1/C_f$

Alternate' Frequency Compensation

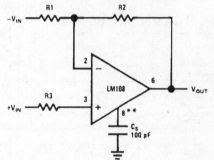

TL/H/7758–2

*Improves rejection of power supply noise by a factor of ten.

**Bandwidth and slew rate are proportional to $1/C_s$

Feedforward Compensation

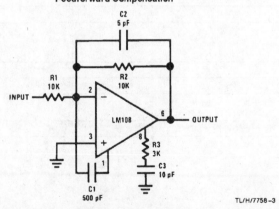

TL/H/7758–3

Absolute Maximum Ratings

If Military/Aerospace specified devices are required, please contact the National Semiconductor Sales Office/ Distributors for availability and specifications.
(Note 5)

	LM108/LM208	LM308
Supply Voltage	$\pm 20V$	$\pm 18V$
Power Dissipation (Note 1)	500 mW	500 mW
Differential Input Current (Note 2)	$\pm 10\,mA$	$\pm 10\,mA$
Input Voltage (Note 3)	$\pm 15V$	$\pm 15V$
Output Short-Circuit Duration	Continuous	Continuous
Operating Temperature Range (LM108)	$-55°C$ to $+125°C$	$0°C$ to $+70°C$
(LM208)	$-25°C$ to $+85°C$	
Storage Temperature Range	$-65°C$ to $+150°C$	$-65°C$ to $+150°C$
Lead Temperature (Soldering, 10 sec)		
DIP	$260°C$	$260°C$
H Package Lead Temp		
(Soldering 10 seconds)	$300°C$	$300°C$
Soldering Information		
Dual-In-Line Package		
Soldering (10 seconds)	$260°C$	
Small Outline Package		
Vapor Phase (60 seconds)	$215°C$	
Infrared (15 seconds)	$220°C$	

See AN-450 "Surface Mounting Methods and Their Effect on Product Reliability" for other methods of soldering surface mount devices.

ESD Tolerance (Note 6)   2000V

Electrical Characteristics (Note 4)

Parameter	Condition	LM108/LM208			LM308			Units
		Min	Typ	Max	Min	Typ	Max	
Input Offset Voltage	$T_A = 25°C$		0.7	2.0		2.0	7.5	mV
Input Offset Current	$T_A = 25°C$		0.05	0.2		0.2	1	nA
Input Bias Current	$T_A = 25°C$		0.8	2.0		1.5	7	nA
Input Resistance	$T_A = 25°C$	30	70		10	40		$M\Omega$
Supply Current	$T_A = 25°C$		0.3	0.6		0.3	0.8	mA
Large Signal Voltage Gain	$T_A = 25°C$, $V_S = \pm15V$ $V_{OUT} = \pm10V$, $R_L \geq 10\,k\Omega$	50	300		25	300		V/mV
Input Offset Voltage				3.0			10	mV
Average Temperature Coefficient of Input Offset Voltage			3.0	15		6.0	30	$\mu V/°C$
Input Offset Current				0.4			1.5	nA
Average Temperature Coefficient of Input Offset Current			0.5	2.5		2.0	10	pA/°C
Input Bias Current				3.0			10	nA
Supply Current	$T_A = +125°C$		0.15	0.4				mA
Large Signal Voltage Gain	$V_S = \pm15V$, $V_{OUT} = \pm10V$ $R_L \geq 10\,k\Omega$	25			15			V/mV
Output Voltage Swing	$V_S = \pm15V$, $R_L = 10\,k\Omega$	±13	±14		±13	±14		V

Typical Performance Characteristics LM108/LM208

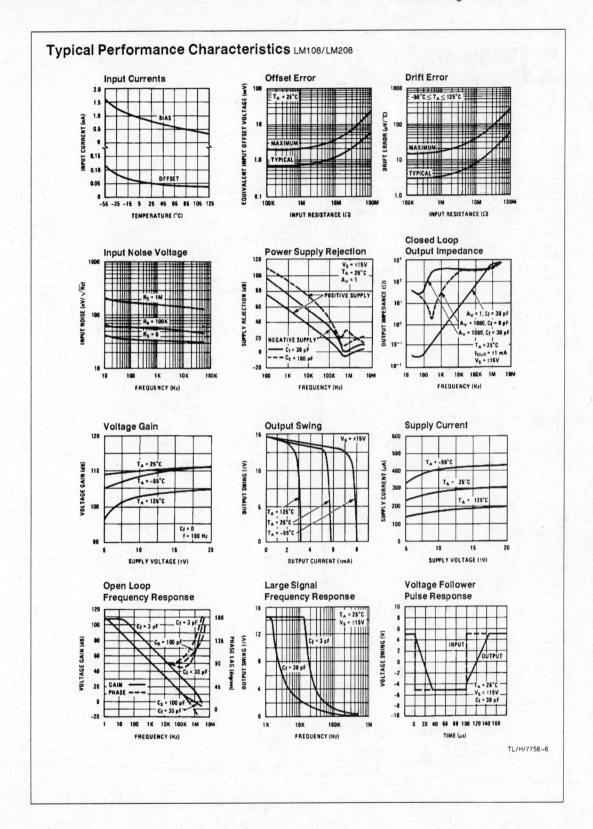

TL/H/7758–6

June 1989

LM111/LM211/LM311 Voltage Comparator

General Description

The LM111, LM211 and LM311 are voltage comparators that have input currents nearly a thousand times lower than devices like the LM106 or LM710. They are also designed to operate over a wider range of supply voltages: from standard ±15V op amp supplies down to the single 5V supply used for IC logic. Their output is compatible with RTL, DTL and TTL as well as MOS circuits. Further, they can drive lamps or relays, switching voltages up to 50V at currents as high as 50 mA.

Both the inputs and the outputs of the LM111, LM211 or the LM311 can be isolated from system ground, and the output can drive loads referred to ground, the positive supply or the negative supply. Offset balancing and strobe capability are provided and outputs can be wire OR'ed. Although slower than the LM106 and LM710 (200 ns response time vs

40 ns) the devices are also much less prone to spurious oscillations. The LM111 has the same pin configuration as the LM106 and LM710.

The LM211 is identical to the LM111, except that its performance is specified over a −25°C to +85°C temperature range instead of −55°C to +125°C. The LM311 has a temperature range of 0°C to +70°C.

Features

- Operates from single 5V supply
- Input current: 150 nA max. over temperature
- Offset current: 20 nA max. over temperature
- Differential input voltage range: ±30V
- Power consumption: 135 mW at ±15V

Typical Applications**

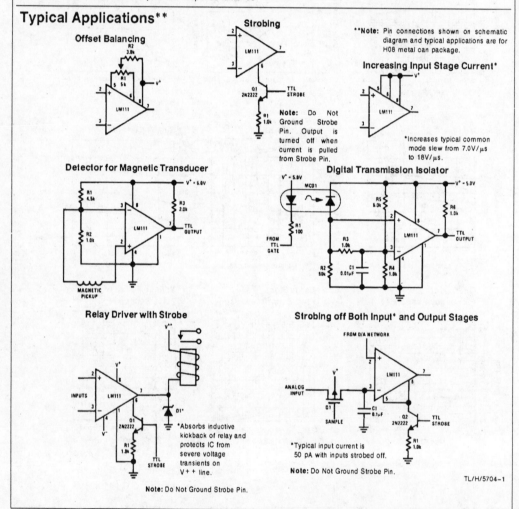

TL/H/5704–1

Reprinted with permission of National Semiconductor Corporation.

Absolute Maximum Ratings for the LM111/LM211

If Military/Aerospace specified devices are required, please contact the National Semiconductor Sales Office/Distributors for availability and specifications. (Note 7)

Total Supply Voltage (V_{84})	36V
Output to Negative Supply Voltage (V_{74})	50V
Ground to Negative Supply Voltage (V_{14})	30V
Differential Input Voltage	$\pm 30V$
Input Voltage (Note 1)	$\pm 15V$
Power Dissipation (Note 2)	500 mW
Output Short Circuit Duration	10 sec

Operating Temperature Range LM111		$-55°C$ to $125°C$
	LM211	$-25°C$ to $85°C$
Storage Temperature Range		$-65°C$ to $150°C$
Lead Temperature (Soldering, 10 sec)		$260°C$
Voltage at Strobe Pin		$V^+ - 5V$

Soldering Information
Dual-In-Line Package
  Soldering (10 seconds) .... $260°C$
Small Outline Package
  Vapor Phase (60 seconds) .... $215°C$
  Infrared (15 seconds) .... $220°C$
See AN-450 "Surface Mounting Methods and Their Effect on Product Reliability" for other methods of soldering surface mount devices.

ESD Rating (Note 8)   300V

Electrical Characteristics for the LM111 and LM211 (Note 3)

Parameter	Conditions	Min	Typ	Max	Units
Input Offset Voltage (Note 4)	$T_A = 25°C$, $R_S \le 50k$		0.7	3.0	mV
Input Offset Current (Note 4)	$T_A = 25°C$		4.0	10	nA
Input Bias Current	$T_A = 25°C$		60	100	nA
Voltage Gain	$T_A = 25°C$	40	200		V/mV
Response Time (Note 5)	$T_A = 25°C$		200		ns
Saturation Voltage	$V_{IN} \le -5$ mV, $I_{OUT} = 50$ mA $T_A = 25°C$		0.75	1.5	V
Strobe ON Current (Note 6)	$T_A = 25°C$	2.0	3.0	5.0	mA
Output Leakage Current	$V_{IN} \ge 5$ mV, $V_{OUT} = 35V$ $T_A = 25°C$, $I_{STROBE} = 3$ mA		0.2	10	nA
Input Offset Voltage (Note 4)	$R_S \le 50$ k			4.0	mV
Input Offset Current (Note 4)				20	nA
Input Bias Current				150	nA
Input Voltage Range	$V^+ = 15V$, $V^- = -15V$, Pin 7 Pull-Up May Go To 5V	-14.5	13.8,-14.7	13.0	V
Saturation Voltage	$V^+ \ge 4.5V$, $V^- = 0$ $V_{IN} \le -6$ mV, $I_{OUT} \le 8$ mA		0.23	0.4	V
Output Leakage Current	$V_{IN} \ge 5$ mV, $V_{OUT} = 35V$		0.1	0.5	μA
Positive Supply Current	$T_A = 25°C$		5.1	6.0	mA
Negative Supply Current	$T_A = 25°C$		4.1	5.0	mA

Note 1: This rating applies for ± 15 supplies. The positive input voltage limit is 30V above the negative supply. The negative input voltage limit is equal to the negative supply voltage or 30V below the positive supply, whichever is less.

Note 2: The maximum junction temperature of the LM111 is 150°C, while that of the LM211 is 110°C. For operating at elevated temperatures, devices in the H08 package must be derated based on a thermal resistance of 165°C/W, junction to ambient, or 20°C/W, junction to case. The thermal resistance of the dual-in-line package is 110°C/W, junction to ambient.

Note 3: These specifications apply for $V_S = \pm 15V$ and Ground pin at ground, and $-55°C \le T_A \le +125°C$, unless otherwise stated. With the LM211, however, all temperature specifications are limited to $-25°C \le T_A \le +85°C$. The offset voltage, offset current and bias current specifications apply for any supply voltage from a single 5V supply up to $\pm 15V$ supplies.

Note 4: The offset voltages and offset currents given are the maximum values required to drive the output within a volt of either supply with a 1 mA load. Thus, these parameters define an error band and take into account the worst-case effects of voltage gain and input impedance.

Note 5: The response time specified (see definitions) is for a 100 mV input step with 5 mV overdrive.

Note 6: This specification gives the range of current which must be drawn from the strobe pin to ensure the output is properly disabled. Do not short the strobe pin to ground; it should be current driven at 3 to 5 mA.

Note 7: Refer to RETS111X for the LM111H, LM111J and LM111J-8 military specifications.

Note 8: Human body model, 1.5 kΩ in series with 100 pF.

Absolute Maximum Ratings for the LM311

If Military/Aerospace specified devices are required, please contact the National Semiconductor Sales Office/Distributors for availability and specifications.

Total Supply Voltage (V_{84})	36V
Output to Negative Supply Voltage V_{74}	40V
Ground to Negative Supply Voltage V_{14})	30V
Differential Input Voltage	± 30V
Input Voltage (Note 1)	± 15V
Power Dissipation (Note 2)	500 mW
ESD Rating (Note 7)	300V

Output Short Circuit Duration	10 sec
Operating Temperature Range	0° to 70°C
Storage Temperature Range	− 65°C to 150°C
Lead Temperature (soldering, 10 sec)	260°C
Voltage at Strobe Pin	$V^+ - 5V$

Soldering Information
Dual-In-Line Package
Soldering (10 seconds) . . . . .260°C
Small Outline Package
Vapor Phase (60 seconds) .215°C
Infrared (15 seconds) . . . . .220°C

See AN-450 "Surface Mounting Methods and Their Effect on Product Reliability" for other methods of soldering surface mount devices.

Electrical Characteristics for the LM311 (Note 3)

Parameter	Conditions	Min	Typ	Max	Units
Input Offset Voltage (Note 4)	$T_A = 25°C$, $R_S \leq 50k$		2.0	7.5	mV
Input Offset Current (Note 4)	$T_A = 25°C$		6.0	50	nA
Input Bias Current	$T_A = 25°C$		100	250	nA
Voltage Gain	$T_A = 25°C$	40	200		V/mV
Response Time (Note 5)	$T_A = 25°C$		200		ns
Saturation Voltage	$V_{IN} \leq - 10$ mV, $I_{OUT} = 50$ mA $T_A = 25°C$		0.75	1.5	V
Strobe ON Current	$T_A = 25°C$	1.5	3.0		mA
Output Leakage Current	$V_{IN} \geq 10$ mV, $V_{OUT} = 35V$ $T_A = 25°C$, $I_{STROBE} = 3$ mA $V^- = V_{GRND} = -5V$		0.2	50	nA
Input Offset Voltage (Note 4)	$R_S \leq 50K$			10	mV
Input Offset Current (Note 4)				70	nA
Input Bias Current				300	nA
Input Voltage Range		−14.5	13.8, − 14.7	13.0	V
Saturation Voltage	$V^+ \geq 4.5V$, $V^- = 0$ $V_{IN} \leq - 10$ mV, $I_{OUT} \leq 8$ mA		0.23	0.4	V
Positive Supply Current	$T_A = 25°C$		5.1	7.5	mA
Negative Supply Current	$T_A = 25°C$		4.1	5.0	mA

Note 1: This rating applies for ± 15V supplies. The positive input voltage limit is 30V above the negative supply. The negative input voltage limit is equal to the negative supply voltage or 30V below the positive supply, whichever is less.

Note 2: The maximum junction temperature of the LM311 is 110°C. For operating at elevated temperature, devices in the H08 package must be derated based on a thermal resistance of 165°C/W, junction to ambient, or 20°C/W, junction to case. The thermal resistance of the dual-in-line package is 100°C/W, junction to ambient.

Note 3: These specifications apply for $V_S = \pm 15V$ and the Ground pin at ground, and 0°C < T_A < + 70°C, unless otherwise specified. The offset voltage, offset current and bias current specifications apply for any supply voltage from a single 5V supply up to ± 15V supplies.

Note 4: The offset voltages and offset currents given are the maximum values required to drive the output within a volt of either supply with 1 mA load. Thus, these parameters define an error band and take into account the worst-case effects of voltage gain and input impedance.

Note 5: The response time specified (see definitions) is for a 100 mV input step with 5 mV overdrive.

Note 6: This specification gives the range of current which must be drawn from the strobe pin to ensure the output is properly disabled. Do not short the strobe pin to ground; it should be current driven at 3 to 5 mA.

Note 7: Human body model, 1.5 kΩ in series with 100 pF.

June 1989

LM117A/LM117/LM317A/LM317
3-Terminal Adjustable Regulator

General Description

The LM117 series of adjustable 3-terminal positive voltage regulators is capable of supplying in excess of 1.5A over a 1.2V to 37V output range. They are exceptionally easy to use and require only two external resistors to set the output voltage. Further, both line and load regulation are better than standard fixed regulators. Also, the LM117 is packaged in standard transistor packages which are easily mounted and handled.

In addition to higher performance than fixed regulators, the LM117 series offers full overload protection available only in IC's. Included on the chip are current limit, thermal overload protection and safe area protection. All overload protection circuitry remains fully functional even if the adjustment terminal is disconnected.

Normally, no capacitors are needed unless the device is situated more than 6 inches from the input filter capacitors in which case an input bypass is needed. An optional output capacitor can be added to improve transient response. The adjustment terminal can be bypassed to achieve very high ripple rejection ratios which are difficult to achieve with standard 3-terminal regulators.

Besides replacing fixed regulators, the LM117 is useful in a wide variety of other applications. Since the regulator is "floating" and sees only the input-to-output differential voltage, supplies of several hundred volts can be regulated as long as the maximum input to output differential is not exceeded, i.e., avoid short-circuiting the output.

Also, it makes an especially simple adjustable switching regulator, a programmable output regulator, or by connecting a fixed resistor between the adjustment pin and output, the LM117 can be used as a precision current regulator. Supplies with electronic shutdown can be achieved by clamping the adjustment terminal to ground which programs the output to 1.2V where most loads draw little current.

The LM117 series devices with a "K" suffix are packaged in standard TO-3 transistor packages, while those with an "H" suffix are in a solid Kovar-base TO-39 transistor package. The LM117A and LM117 are rated for operation from $-55°C$ to $+150°C$, the LM317A from $-40°C$ to $+125°C$, and the LM317 from 0°C to $+125°C$. The LM317AT and the LM317T are available in a TO-220 plastic package and the LM317MP in a TO-202 plastic package.

For applications requiring greater output current, see LM150 series (3A) and LM138 series (5A) data sheets. For the negative complement, see LM137 series data sheet.

LM117 Series Packages and Power Capability

Part Number Suffix	Package	Rated Power Dissipation	Design Load Current
K	TO-3	20W	1.5A
H	TO-39	2W	0.5A
T	TO-220	20W	1.5A
MP	TO-202	2W	0.5A

Features

- Guaranteed 1% output voltage tolerance (LM117A, LM317A)
- Guaranteed max. 0.01%/V line regulation (LM117A, LM317A)
- Guaranteed max. 0.3% load regulation (LM117A, LM117)
- Guaranteed 1.5A output current
- Adjustable output down to 1.2V
- Current limit constant with temperature
- 100% electrical burn-in
- 80 dB ripple rejection
- Output is short-circuit protected

Typical Applications

1.2V–25V Adjustable Regulator

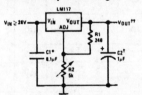

TL/H/9063–1

Full output current not available at high input-output voltages

*Needed if device is more than 6 inches from filter capacitors.

†Optional—improves transient response. Output capacitors in the range of 1 µF to 1000 µF of aluminum or tantalum electrolytic are commonly used to provide improved output impedance and rejection of transients.

$$††V_{OUT} = 1.25V\left(1 + \frac{R2}{R1}\right) + I_{ADJ}(R_2)$$

Digitally Selected Outputs

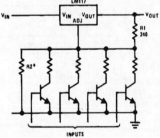

TL/H/9063–2

*Sets maximum V_{OUT}

Reprinted with permission of National Semiconductor Corporation.

Absolute Maximum Ratings (Note 1)

If Military/Aerospace specified devices are required, please contact the National Semiconductor Sales Office/Distributors for availability and specifications. (Note 2)

Power Dissipation	Internally Limited
Input-Output Voltage Differential	$+40V, -0.3V$
Storage Temperature	$-65°C$ to $+150°C$
Lead Temperature	
Metal Package (Soldering, 10 seconds)	300°C
Plastic Package (Soldering, 4 seconds)	260°C
ESD Tolerance (Note 5)	3 kV

Operating Temperature Range

LM117A/LM117	$-55°C \leq T_J \leq +150°C$
LM317A	$-40°C \leq T_J \leq +125°C$
LM317	$0°C \leq T_J \leq +125°C$

Preconditioning

Thermal Limit Burn-In	All Devices 100%

Electrical Characteristics

Specifications with standard type face are for $T_J = 25°C$, and those with **boldface type** apply over full **Operating Temperature Range**. Unless otherwise specified, $V_{IN} - V_{OUT} = 5V$, and $I_{OUT} = 10$ mA. (Note 3)

Parameter	Conditions	LM117A			LM117			Units
		Min	Typ	Max	Min	Typ	Max	
Reference Voltage		1.238	1.250	1.262				V
	$3V \leq (V_{IN} - V_{OUT}) \leq 40V$, 10 mA $\leq I_{OUT} \leq I_{MAX}, P \leq P_{MAX}$	**1.225**	**1.250**	**1.270**	**1.20**	**1.25**	**1.30**	V
Line Regulation	$3V \leq (V_{IN} - V_{OUT}) \leq 40V$ (Note 4)		0.005	0.01		0.01	0.02	%/V
			0.01	**0.02**		**0.02**	**0.05**	%/V
Load Regulation	10 mA $\leq I_{OUT} \leq I_{MAX}$ (Note 4)		0.1	0.3		0.1	0.3	%
			0.3	**1**		**0.3**	**1**	%
Thermal Regulation	20 ms Pulse		0.03	0.07		0.03	0.07	%/W
Adjustment Pin Current			**50**	**100**		**50**	**100**	µA
Adjustment Pin Current Change	10 mA $\leq I_{OUT} \leq I_{MAX}$ $3V \leq (V_{IN} - V_{OUT}) \leq 40V$		**0.2**	**5**		**0.2**	**5**	µA
Temperature Stability	$T_{MIN} \leq T_J \leq T_{MAX}$		**1**			**1**		%
Minimum Load Current	$(V_{IN} - V_{OUT}) = 40V$		**3.5**	**5**		**3.5**	**5**	mA
Current Limit	$(V_{IN} - V_{OUT}) \leq 15V$ K Package H Package	**1.5** **0.5**	**2.2** **0.8**	**3.4** **1.8**	**1.5** **0.5**	**2.2** **0.8**	**3.4** **1.8**	A A
	$(V_{IN} - V_{OUT}) = 40V$ K Package H Package	0.3 0.15	0.4 0.2		0.3 0.15	0.4 0.2		A A
RMS Output Noise, % of V_{OUT}	10 Hz $\leq f \leq 10$ kHz		0.003			0.003		%
Ripple Rejection Ratio	$V_{OUT} = 10V, f = 120$ Hz, $C_{ADJ} = 0$ µF		**65**			**65**		dB
	$V_{OUT} = 10V, f = 120$ Hz, $C_{ADJ} = 10$ µF	**66**	**80**		**66**	**80**		dB
Long-Term Stability	$T_J = 125°C, 1000$ hrs		0.3	1		0.3	1	%
Thermal Resistance, Junction-to-Case	K Package H Package		2.3 12	3 15		2.3 12	3 15	°C/W °C/W
Thermal Resistance, Junction-to-Ambient (No Heat Sink)	K Package H Package		35 140			35 140		°C/W °C/W

Electrical Characteristics (Continued)

Specifications with standard type face are for T_J = 25°C, and those with **boldface type** apply over **full Operating Temperature Range**. Unless otherwise specified, $V_{IN} - V_{OUT}$ = 5V, and I_{OUT} = 10 mA. (Note 3)

Parameter	Conditions	LM317A			LM317			Units
		Min	Typ	Max	Min	Typ	Max	
Reference Voltage		1.238	1.250	1.262				V
	$3V \leq (V_{IN} - V_{OUT}) \leq 40V$, $10\,mA \leq I_{OUT} \leq I_{MAX}, P \leq P_{MAX}$	**1.225**	**1.250**	**1.270**	**1.20**	**1.25**	**1.30**	V
Line Regulation	$3V \leq (V_{IN} - V_{OUT}) \leq 40V$ (Note 4)		0.005	0.01		0.01	0.04	%/V
			0.01	**0.02**		**0.02**	**0.07**	%/V
Load Regulation	$10\,mA \leq I_{OUT} \leq I_{MAX}$ (Note 4)		0.1	0.5		0.1	0.5	%
			0.3	**1**		**0.3**	**1.5**	%
Thermal Regulation	20 ms Pulse		0.04	0.07		0.04	0.07	%/W
Adjustment Pin Current			**50**	**100**		**50**	**100**	µA
Adjustment Pin Current Change	$10\,mA \leq I_{OUT} \leq I_{MAX}$ $3V \leq (V_{IN} - V_{OUT}) \leq 40V$		**0.2**	**5**		**0.2**	**5**	µA
Temperature Stability	$T_{MIN} \leq T_J \leq T_{MAX}$		**1**			**1**		%
Minimum Load Current	$(V_{IN} - V_{OUT})$ = 40V		**3.5**	**10**		**3.5**	**10**	mA
Current Limit	$(V_{IN} - V_{OUT}) \leq 15V$ K and T Package H Package P Package	**1.5** **0.5**	**2.2** **0.8**	**3.4** **1.8**	**1.5** **0.5** **0.5**	**2.2** **0.8** **0.8**	**3.4** **1.8** **1.8**	A A A
	$(V_{IN} - V_{OUT})$ = 40V K and T Package H Package P Package	0.15 0.075	0.4 0.2		0.15 0.075 0.075	0.4 0.2 0.2		A A A
RMS Output Noise, % of V_{OUT}	$10\,Hz \leq f \leq 10\,kHz$		0.003			0.003		%
Ripple Rejection Ratio	V_{OUT} = 10V, f = 120 Hz, C_{ADJ} = 0 µF		**65**			**65**		dB
	V_{OUT} = 10V, f = 120 Hz, C_{ADJ} = 10 µF	**66**	**80**		**66**	**80**		dB
Long-Term Stability	T_J = 125°C, 1000 hrs		0.3	1		0.3	1	%
Thermal Resistance, Junction-to-Case	K Package H Package T Package P Package		2.3 12 4	3 15 5		2.3 12 4 7	3 15	°C/W °C/W °C/W °C/W
Thermal Resistance, Junction-to-Ambient (No Heat Sink)	K Package H Package T Package P Package		35 140 50			35 140 50 80		°C/W °C/W °C/W °C/W

Note 1: Absolute Maximum Ratings indicate limits beyond which damage to the device may occur. Operating Ratings indicate conditions for which the device is intended to be functional, but do not guarantee specific performance limits. For guaranteed specifications and test conditions, see the Electrical Characteristics. The guaranteed specifications apply only for the test conditions listed.

Note 2: Refer to RETS117AH drawing for the LM117AH, the RETS117H drawing for the LM117H, the RETS117AK drawing for the LM117AK, or the RETS117K for the LM117K military specifications.

Note 3: Although power dissipation is internally limited, these specifications are applicable for maximum power dissipations of 2W for the TO-39 and TO-202, and 20W for the TO-3 and TO-220. I_{MAX} is 1.5A for the TO-3 and TO-220 packages and 0.5A for the TO-39 and TO-202 packages. All limits (i.e., the numbers in the Min. and Max. columns) are guaranteed to National's AOQL (Average Outgoing Quality Level).

Note 4: Regulation is measured at a constant junction temperature, using pulse testing with a low duty cycle. Changes in output voltage due to heating effects are covered under the specifications for thermal regulation.

Note 5: Human body model, 100 pF discharged through a 1.5 kΩ resistor.

April 1989

LM555/LM555C Timer

General Description

The LM555 is a highly stable device for generating accurate time delays or oscillation. Additional terminals are provided for triggering or resetting if desired. In the time delay mode of operation, the time is precisely controlled by one external resistor and capacitor. For astable operation as an oscillator, the free running frequency and duty cycle are accurately controlled with two external resistors and one capacitor. The circuit may be triggered and reset on falling waveforms, and the output circuit can source or sink up to 200 mA or drive TTL circuits.

Features

- Direct replacement for SE555/NE555
- Timing from microseconds through hours
- Operates in both astable and monostable modes

- Adjustable duty cycle
- Output can source or sink 200 mA
- Output and supply TTL compatible
- Temperature stability better than 0.005% per °C
- Normally on and normally off output

Applications

- Precision timing
- Pulse generation
- Sequential timing
- Time delay generation
- Pulse width modulation
- Pulse position modulation
- Linear ramp generator

Schematic Diagram

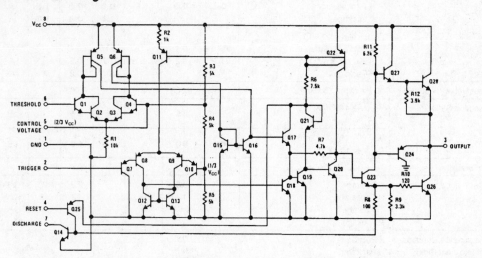

TL/H/7851-1

Absolute Maximum Ratings

If Military/Aerospace specified devices are required, please contact the National Semiconductor Sales Office/Distributors for availability and specifications.

Supply Voltage $+18V$

Power Dissipation (Note 1)   LM555H, LM555CH 760 mW

Operating Temperature Ranges

LM555N, LM555CN 1180 mW

LM555C $0°C$ to $+70°C$

LM555 $-55°C$ to $+125°C$

Storage Temperature Range $-65°C$ to $+150°C$

Soldering Information
Dual-In-Line Package
Soldering (10 Seconds) $260°C$
Small Outline Package
Vapor Phase (60 Seconds) $215°C$
Infrared (15 Seconds) $220°C$

See AN-450 "Surface Mounting Methods and Their Effect on Product Reliability" for other methods of soldering surface mount devices.

Electrical Characteristics ($T_A = 25°C$, $V_{CC} = +5V$ to $+15V$, unless othewise specified)

Parameter	Conditions	Limits						Units
		LM555			LM555C			
		Min	Typ	Max	Min	Typ	Max	
Supply Voltage		4.5		18	4.5		16	V
Supply Current	$V_{CC} = 5V$, $R_L = \infty$		3	5		3	6	mA
	$V_{CC} = 15V$, $R_L = \infty$ (Low State) (Note 2)		10	12		10	15	mA
Timing Error, Monostable Initial Accuracy Drift with Temperature	$R_A = 1k$ to $100\ k\Omega$, $C = 0.1\ \mu F$, (Note 3)		0.5 30			1 50		% ppm/°C
Accuracy over Temperature Drift with Supply			1.5 0.05			1.5 0.1		% %/V
Timing Error, Astable Initial Accuracy Drift with Temperature	R_A, $R_B = 1k$ to $100\ k\Omega$, $C = 0.1\ \mu F$, (Note 3)		1.5 90			2.25 150		% ppm/°C
Accuracy over Temperature Drift with Supply			2.5 0.15			3.0 0.30		% %/V
Threshold Voltage			0.667			0.667		x V_{CC}
Trigger Voltage	$V_{CC} = 15V$ $V_{CC} = 5V$	4.8 1.45	5 1.67	5.2 1.9		5 1.67		V V
Trigger Current			0.01	0.5		0.5	0.9	μA
Reset Voltage		0.4	0.5	1	0.4	0.5	1	V
Reset Current			0.1	0.4		0.1	0.4	mA
Threshold Current	(Note 4)		0.1	0.25		0.1	0.25	μA
Control Voltage Level	$V_{CC} = 15V$ $V_{CC} = 5V$	9.6 2.9	10 3.33	10.4 3.8	9 2.6	10 3.33	11 4	V V
Pin 7 Leakage Output High			1	100		1	100	nA
Pin 7 Sat (Note 5) Output Low Output Low	$V_{CC} = 15V$, $I_7 = 15\ mA$ $V_{CC} = 4.5V$, $I_7 = 4.5\ mA$		150 70	100		180 80	200	mV mV

Electrical Characteristics T_A = 25°C, V_{CC} = +5V to +15V, (unless othewise specified) (Continued)

Parameter	Conditions	Limits						Units
		LM555			LM555C			
		Min	Typ	Max	Min	Typ	Max	
Output Voltage Drop (Low)	V_{CC} = 15V							
	I_{SINK} = 10 mA		0.1	0.15		0.1	0.25	V
	I_{SINK} = 50 mA		0.4	0.5		0.4	0.75	V
	I_{SINK} = 100 mA		2	2.2		2	2.5	V
	I_{SINK} = 200 mA		2.5			2.5		V
	V_{CC} = 5V							
	I_{SINK} = 8 mA		0.1	0.25				V
	I_{SINK} = 5 mA					0.25	0.35	V
Output Voltage Drop (High)	I_{SOURCE} = 200 mA, V_{CC} = 15V		12.5			12.5		V
	I_{SOURCE} = 100 mA, V_{CC} = 15V	13	13.3		12.75	13.3		V
	V_{CC} = 5V	3	3.3		2.75	3.3		V
Rise Time of Output			100			100		ns
Fall Time of Output			100			100		ns

Note 1: For operating at elevated temperatures the device must be derated above 25°C based on a +150°C maximum junction temperature and a thermal resistance of 164°c/w (TO-5), 106°c/w (DIP) and 170°c/w (SO-8) junction to ambient.

Note 2: Supply current when output high typically 1 mA less at V_{CC} = 5V.

Note 3: Tested at V_{CC} = 5V and V_{CC} = 15V.

Note 4: This will determine the maximum value of R_A + R_B for 15V operation. The maximum total (R_A + R_B) is 20 MΩ.

Note 5: No protection against excessive pin 7 current is necessary providing the package dissipation rating will not be exceeded.

Note 6: Refer to RETS555X drawing of military LM555H and LM555J versions for specifications.

Connection Diagrams

Metal Can Package

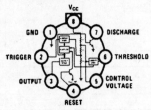

TL/H/7851–2

Top View

Order Number LM555H or LM555CH
See NS Package Number H08C

Dual-In-Line and Small Outline Packages

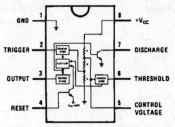

TL/H/7851–3

Top View

Order Number LM555J, LM555CJ,
LM555CM or LM555CN
See NS Package Number J08A, M08A or N08E

May 1989

LM741/LM741A/LM741C/LM741E Operational Amplifier

General Description

The LM741 series are general purpose operational amplifiers which feature improved performance over industry standards like the LM709. They are direct, plug-in replacements for the 709C, LM201, MC1439 and 748 in most applications.

The amplifiers offer many features which make their application nearly foolproof: overload protection on the input and output, no latch-up when the common mode range is exceeded, as well as freedom from oscillations.

The LM741C/LM741E are identical to the LM741/LM741A except that the LM741C/LM741E have their performance guaranteed over a 0°C to +70°C temperature range, instead of −55°C to +125°C.

Schematic and Connection Diagrams (Top Views)

TL/H/9341–1

Metal Can Package

NC

TL/H/9341–2

Order Number LM741H, LM741AH, LM741CH or LM741EH
See NS Package Number H08C

Dual-In-Line or S.O. Package

OFFSET NULL	1	8	NC
INVERTING INPUT	2	7	V⁺
NON-INVERTING INPUT	3	6	OUTPUT
V⁻	4	5	OFFSET NULL

TL/H/9341–3

Order Number LM741J, LM741AJ, LM741CJ, LM741CM, LM741CN or LM741EN
See NS Package Number J08A, M08A or N08E

Absolute Maximum Ratings

If Military/Aerospace specified devices are required, please contact the National Semiconductor Sales Office/Distributors for availability and specifications. (Note 5)

	LM741A	LM741E	LM741	LM741C
Supply Voltage	$\pm 22V$	$\pm 22V$	$\pm 22V$	$\pm 18V$
Power Dissipation (Note 1)	500 mW	500 mW	500 mW	500 mW
Differential Input Voltage	$\pm 30V$	$\pm 30V$	$\pm 30V$	$\pm 30V$
Input Voltage (Note 2)	$\pm 15V$	$\pm 15V$	$\pm 15V$	$\pm 15V$
Output Short Circuit Duration	Continuous	Continuous	Continuous	Continuous
Operating Temperature Range	$-55°C$ to $+125°C$	$0°C$ to $+70°C$	$-55°C$ to $+125°C$	$0°C$ to $+70°C$
Storage Temperature Range	$-65°C$ to $+150°C$	$-65°C$ to $+150°C$	$-65°C$ to $+150°C$	$-65°C$ to $+150°C$
Junction Temperature	$150°C$	$100°C$	$150°C$	$100°C$
Soldering Information				
N-Package (10 seconds)	$260°C$	$260°C$	$260°C$	$260°C$
J- or H-Package (10 seconds)	$300°C$	$300°C$	$300°C$	$300°C$
M-Package				
Vapor Phase (60 seconds)	$215°C$	$215°C$	$215°C$	$215°C$
Infrared (15 seconds)	$215°C$	$215°C$	$215°C$	$215°C$

See AN-450 "Surface Mounting Methods and Their Effect on Product Reliability" for other methods of soldering surface mount devices.

ESD Tolerance (Note 6)	400V	400V	400V	400V

Electrical Characteristics (Note 3)

Parameter	Conditions	LM741A/LM741E			LM741			LM741C			Units
		Min	Typ	Max	Min	Typ	Max	Min	Typ	Max	
Input Offset Voltage	$T_A = 25°C$										
	$R_S \leq 10\,k\Omega$					1.0	5.0		2.0	6.0	mV
	$R_S \leq 50\,\Omega$		0.8	3.0							mV
	$T_{AMIN} \leq T_A \leq T_{AMAX}$										
	$R_S \leq 50\,\Omega$			4.0							mV
	$R_S \leq 10\,k\Omega$						6.0			7.5	mV
Average Input Offset Voltage Drift				15							$\mu V/°C$
Input Offset Voltage Adjustment Range	$T_A = 25°C, V_S = \pm 20V$	± 10				± 15			± 15		mV
Input Offset Current	$T_A = 25°C$		3.0	30		20	200		20	200	nA
	$T_{AMIN} \leq T_A \leq T_{AMAX}$			70		85	500			300	nA
Average Input Offset Current Drift				0.5							$nA/°C$
Input Bias Current	$T_A = 25°C$		30	80		80	500		80	500	nA
	$T_{AMIN} \leq T_A \leq T_{AMAX}$			0.210			1.5			0.8	μA
Input Resistance	$T_A = 25°C, V_S = \pm 20V$	1.0	6.0		0.3	2.0		0.3	2.0		$M\Omega$
	$T_{AMIN} \leq T_A \leq T_{AMAX}$, $V_S = \pm 20V$	0.5									$M\Omega$
Input Voltage Range	$T_A = 25°C$							± 12	± 13		V
	$T_{AMIN} \leq T_A \leq T_{AMAX}$				± 12	± 13					V
Large Signal Voltage Gain	$T_A = 25°C, R_L \geq 2\,k\Omega$										
	$V_S = \pm 20V, V_O = \pm 15V$	50									V/mV
	$V_S = \pm 15V, V_O = \pm 10V$				50	200		20	200		V/mV
	$T_{AMIN} \leq T_A \leq T_{AMAX}$, $R_L \geq 2\,k\Omega$,										
	$V_S = \pm 20V, V_O = \pm 15V$	32									V/mV
	$V_S = \pm 15V, V_O = \pm 10V$				25			15			V/mV
	$V_S = \pm 5V, V_O = \pm 2V$	10									V/mV

Electrical Characteristics (Note 3) (Continued)

Parameter	Conditions	LM741A/LM741E			LM741			LM741C			Units
		Min	Typ	Max	Min	Typ	Max	Min	Typ	Max	
Output Voltage Swing	$V_S = \pm 20V$ $R_L \geq 10\,k\Omega$ $R_L \geq 2\,k\Omega$	± 16 ± 15									V V
	$V_S = \pm 15V$ $R_L \geq 10\,k\Omega$ $R_L \geq 2\,k\Omega$				± 12 ± 10	± 14 ± 13		± 12 ± 10	± 14 ± 13		V V
Output Short Circuit Current	$T_A = 25°C$ $T_{AMIN} \leq T_A \leq T_{AMAX}$	10 10	25	35 40		25			25		mA mA
Common-Mode Rejection Ratio	$T_{AMIN} \leq T_A \leq T_{AMAX}$ $R_S \leq 10\,k\Omega, V_{CM} = \pm 12V$ $R_S \leq 50\Omega, V_{CM} = \pm 12V$	80	95		70	90		70	90		dB dB
Supply Voltage Rejection Ratio	$T_{AMIN} \leq T_A \leq T_{AMAX},$ $V_S = \pm 20V$ to $V_S = \pm 5V$ $R_S \leq 50\Omega$ $R_S \leq 10\,k\Omega$	86	96		77	96		77	96		dB dB
Transient Response Rise Time Overshoot	$T_A = 25°C$, Unity Gain		0.25 6.0	0.8 20		0.3 5			0.3 5		μs %
Bandwidth (Note 4)	$T_A = 25°C$	0.437	1.5								MHz
Slew Rate	$T_A = 25°C$, Unity Gain	0.3	0.7			0.5			0.5		V/μs
Supply Current	$T_A = 25°C$					1.7	2.8		1.7	2.8	mA
Power Consumption	$T_A = 25°C$ $V_S = \pm 20V$ $V_S = \pm 15V$		80	150		50	85		50	85	mW mW
LM741A	$V_S = \pm 20V$ $T_A = T_{AMIN}$ $T_A = T_{AMAX}$			165 135							mW mW
LM741E	$V_S = \pm 20V$ $T_A = T_{AMIN}$ $T_A = T_{AMAX}$			150 150							mW mW
LM741	$V_S = \pm 15V$ $T_A = T_{AMIN}$ $T_A = T_{AMAX}$					60 45	100 75				mW mW

Note 1: For operation at elevated temperatures, these devices must be derated based on thermal resistance, and T_j max. (listed under "Absolute Maximum Ratings"). $T_j = T_A + (\theta_{jA} P_D)$.

Thermal Resistance	Cerdip (J)	DIP (N)	HO8 (H)	SO-8 (M)
θ_{jA} (Junction to Ambient)	100°C/W	100°C/W	170°C/W	195°C/W
θ_{jC} (Junction to Case)	N/A	N/A	25°C/W	N/A

Note 2: For supply voltages less than $\pm 15V$, the absolute maximum input voltage is equal to the supply voltage.

Note 3: Unless otherwise specified, these specifications apply for $V_S = \pm 15V$, $-55°C \leq T_A \leq +125°C$ (LM741/LM741A). For the LM741C/LM741E, these specifications are limited to $0°C \leq T_A \leq +70°C$.

Note 4: Calculated value from: BW (MHz) = 0.35/Rise Time(μs).

Note 5: For military specifications see RETS741X for LM741 and RETS741AX for LM741A.

Note 6: Human body model, 1.5 kΩ in series with 100 pF.

Digital Integrated Circuits

DIGITAL INTEGRATED CIRCUITS - COST DATA

54' RATINGS

Functional Type	54'	54LS'	54AS'	54HC'
54xx00	$1.20	$ .80	$1.40	$1.00
54xx02	1.20	.80	1.40	1.00
54xx04	1.35	.81	1.50	1.00
54xx05	1.90	.81	NA	1.20
54xx08	1.20	.80	1.40	1.00
54xx32	1.20	.80	1.40	1.00
54xx74	1.92	1.08	2.00	1.45
54xx109	2.08	1.07	2.10	1.55

74' RATINGS

Functional Type	74'	74LS'	74AS'	74HC'
74xx00	$ .49	$ .27	$ .44	$ .36
74xx02	.49	.27	.44	.36
74xx04	.51	.27	.44	.36
74xx05	.51	.27	NA	.44
74xx08	.49	.27	.44	.36
74xx32	.49	.27	.44	.36
74xx74	.64	.30	.56	.56
74xx109	.64	.44	.67	.71

Seven-Segment Decoder/Driver

7446A	$1.11
7447A	1.09

Seven-Segment Red LED Displays

AND332S, 333S, 334S	$2.05
AND342S, 343S, 344S	2.30
AND362S, 363S, 364S, 365S	2.94

54/74 LOGIC, Pin Assignments, (Top Views)

00
QUADRUPLE
2-INPUT
NAND GATE

$Y = \overline{AB}$

02
QUADRUPLE
2-INPUT
NOR GATE

$Y = \overline{A+B}$

04
HEX INVERTER

HEX INVERTER
WITH OPEN
COLLECTOR
(DRAIN)

$Y = \overline{A}$

08
QUADRUPLE
2-INPUT
AND GATE

$Y = AB$

32
QUADRUPLE
2-INPUT
OR GATE

$Y = A+B$

46A, 47A
BCD-SEVEN-SEGMENT
DECODERS/DRIVERS

74 DUAL D-TYPE POSITIVE-EDGE TRIGGERED FLIP-FLOPS WITH PRESET AND CLEAR

Function Table

	INPUTS			OUTPUTS	
PRESET	CLEAR	CLOCK	D	Q	$\overline{Q}$
L	H	X	X	H	L
H	L	X	X	L	H
L	L	X	X	H*	H*
H	H	↑	H	H	L
H	H	↑	L	L	H
H	H	L	X	Q	$\overline{Q}$

109 DUAL J-$\overline{K}$ POSITIVE-EDGE TRIGGERED FLIP-FLOPS WITH PRESET AND CLEAR

Function Table

	INPUTS				OUTPUTS	
PRESET	CLEAR	CLOCK	J	$\overline{K}$	Q	$\overline{Q}$
L	H	X	X	X	H	L
H	L	X	X	X	L	H
L	L	X	X	X	H*	H*
H	H	↑	L	L	L	H
H	H	↑	H	L	TOGGLE	
H	H	↑	L	H	Q	$\overline{Q}$
H	H	↑	H	H	H	L
H	H	L	X	X	Q	$\overline{Q}$

*This condition is unstable, it will not persist when the preset and clear inputs return to the high state.

recommended operating conditions

		SN5400			SN7400			UNIT
		MIN	NOM	MAX	MIN	NOM	MAX	
V_{CC}	Supply voltage	4.5	5	5.5	4.75	5	5.25	V
V_{IH}	High-level input voltage	2			2			V
V_{IL}	Low-level input voltage			0.8			0.8	V
I_{OH}	High-level output current			− 0.4			− 0.4	mA
I_{OL}	Low-level output current			16			16	mA
T_A	Operating free-air temperature	− 55		125	0		70	°C

electrical characteristics over recommended operating free-air temperature range (unless otherwise noted)

PARAMETER	TEST CONDITIONS †			SN5400			SN7400			UNIT
				MIN	TYP‡	MAX	MIN	TYP‡	MAX	
V_{IK}	V_{CC} = MIN,	I_I = − 12 mA				− 1.5			− 1.5	V
V_{OH}	V_{CC} = MIN,	V_{IL} = 0.8 V,	I_{OH} = − 0.4 mA	2.4	3.4		2.4	3.4		V
V_{OL}	V_{CC} = MIN,	V_{IH} = 2 V,	I_{OL} = 16 mA		0.2	0.4		0.2	0.4	V
I_I	V_{CC} = MAX,	V_I = 5.5 V				1			1	mA
I_{IH}	V_{CC} = MAX,	V_I = 2.4 V				40			40	µA
I_{IL}	V_{CC} = MAX,	V_I = 0.4 V				− 1.6			− 1.6	mA
I_{OS}§	V_{CC} = MAX			− 20		− 55	− 18		− 55	mA
I_{CCH}	V_{CC} = MAX,	V_I = 0 V			4	8		4	8	mA
I_{CCL}	V_{CC} = MAX,	V_I = 4.5 V			12	22		12	22	mA

† For conditions shown as MIN or MAX, use the appropriate value specified under recommended operating conditions.
‡ All typical values are at V_{CC} = 5 V, T_A = 25°C.
§ Not more than one output should be shorted at a time.

switching characteristics, V_{CC} = 5 V, T_A = 25°C (see note 2)

PARAMETER	FROM (INPUT)	TO (OUTPUT)	TEST CONDITIONS		MIN	TYP	MAX	UNIT
t_{PLH}	A or B	Y	R_L = 400 Ω,	C_L = 15 pF		11	22	ns
t_{PHL}						7	15	ns

NOTE 2: Load circuits and voltage waveforms are shown in Section 1.

TEXAS INSTRUMENTS
POST OFFICE BOX 655012 • DALLAS, TEXAS 75265

TTL Devices

SN5402, SN7402
QUADRUPLE 2-INPUT POSITIVE-NOR GATES

recommended operating conditions

		SN5402			SN7402			UNIT
		MIN	NOM	MAX	MIN	NOM	MAX	
V_{CC}	Supply voltage	4.5	5	5.5	4.75	5	5.25	V
V_{IH}	High-level input voltage	2			2			V
V_{IL}	Low-level input voltage			0.8			0.8	V
I_{OH}	High-level output current			− 0.4			− 0.4	mA
I_{OL}	Low-level output current			16			16	mA
T_A	Operating free-air temperature	− 55		125	0		70	°C

electrical characteristics over recommended operating free-air temperature range (unless otherwise noted)

PARAMETER	TEST CONDITIONS †			SN5402			SN7402			UNIT
			MIN	TYP‡	MAX	MIN	TYP‡	MAX		
V_{IK}	V_{CC} = MIN,	I_I = − 12 mA			− 1.5			− 1.5		V
V_{OH}	V_{CC} = MIN,	V_{IL} = 0.8 V,	I_{OH} = − 0.4 mA	2.4	3.4		2.4	3.4		V
V_{OL}	V_{CC} = MIN,	V_{IH} = 2 V,	I_{OL} = 16 mA		0.2	0.4		0.2	0.4	V
I_I	V_{CC} = MAX,	V_I = 5.5 V			1			1		mA
I_{IH}	V_{CC} = MAX,	V_I = 2.4 V			40			40		µA
I_{IL}	V_{CC} = MAX,	V_I = 0.4 V			− 1.6			− 1.6		mA
I_{OS}§	V_{CC} = MAX		− 20		− 55	− 18		− 55		mA
I_{CCH}	V_{CC} = MAX,	V_I = 0 V		8	16		8	16		mA
I_{CCL}	V_{CC} = MAX,	See Note 2		14	27		14	27		mA

† For conditions shown as MIN or MAX, use the appropriate value specified under recommended operating conditions.
‡ All typical values are at V_{CC} = 5 V, T_A = 25°C.
§ Not more than one output should be shorted at a time.
NOTE 2: One input at 4.5 V, all others at GND.

switching characteristics, V_{CC} = 5 V, T_A = 25°C (see note 3)

PARAMETER	FROM (INPUT)	TO (OUTPUT)	TEST CONDITIONS		MIN	TYP	MAX	UNIT
t_{PLH}	A or B	Y	R_L = 400 Ω,	C_L = 15 pF		12	22	ns
t_{PHL}						8	15	ns

NOTE 3: Load circuits and voltage waveforms are shown in Section 1.

TEXAS
INSTRUMENTS
POST OFFICE BOX 655012 • DALLAS. TEXAS 75265

TTL Devices

TTL Devices

recommended operating conditions

		SN5404			SN7404			UNIT
		MIN	NOM	MAX	MIN	NOM	MAX	
V_{CC}	Supply voltage	4.5	5	5.5	4.75	5	5.25	V
V_{IH}	High-level input voltage	2			2			V
V_{IL}	Low-level input voltage			0.8			0.8	V
I_{OH}	High-level output current			−0.4			−0.4	mA
I_{OL}	Low-level output current			16			16	mA
T_A	Operating free-air temperature	−55		125	0		70	$^\circ$C

electrical characteristics over recommended operating free-air temperature range (unless otherwise noted)

PARAMETER	TEST CONDITIONS †			SN5404			SN7404			UNIT
				MIN	TYP‡	MAX	MIN	TYP‡	MAX	
V_{IK}	V_{CC} MIN,	$I_I = -12$ mA				−1.5			−1.5	V
V_{OH}	V_{CC} = MIN,	V_{IL} = 0.8 V,	$I_{OH} = -0.4$ mA	2.4	3.4		2.4	3.4		V
V_{OL}	V_{CC} = MIN,	V_{IH} = 2 V,	I_{OL} = 16 mA		0.2	0.4		0.2	0.4	V
I_I	V_{CC} = MAX,	V_I = 5.5 V				1			1	mA
I_{IH}	V_{CC} = MAX,	V_I = 2.4 V				40			40	μA
I_{IL}	V_{CC} = MAX,	V_I = 0.4 V				−1.6			−1.6	mA
I_{OS} §	V_{CC} = MAX			−20		−55	−18		−55	mA
I_{CCH}	V_{CC} = MAX,	V_I = 0 V			6	12		6	12	mA
I_{CCL}	V_{CC} = MAX,	V_I = 4.5 V			18	33		18	33	mA

† For conditions shown as MIN or MAX, use the appropriate value specified under recommended operating conditions.
‡ All typical values are at V_{CC} 5 V, T_A 25°C.
§ Not more than one output should be shorted at a time.

switching characteristics, V_{CC} = 5 V, T_A = 25°C (see note 2)

PARAMETER	FROM (INPUT)	TO (OUTPUT)	TEST CONDITIONS		MIN	TYP	MAX	UNIT
t_{PLH}	A	Y	R_L = 400 Ω,	C_L = 15 pF		12	22	ns
t_{PHL}						8	15	ns

NOTE 2: Load circuits and voltage waveforms are shown in Section 1.

TEXAS
INSTRUMENTS
POST OFFICE BOX 655012 • DALLAS, TEXAS 75265

SN5405, SN7405
HEX INVERTERS WITH OPEN-COLLECTOR OUTPUTS

recommended operating conditions

		SN5405			SN7405			UNIT
		MIN	NOM	MAX	MIN	NOM	MAX	
V_{CC}	Supply voltage	4.5	5	5.5	4.75	5	5.25	V
V_{IH}	High-level input voltage	2			2			V
V_{IL}	Low-level input voltage			0.8			0.8	V
V_{OH}	High-level output voltage			5.5			5.5	V
I_{OL}	Low-level output current			16			16	mA
T_A	Operating free-air temperature	−55		125	0		70	°C

electrical characteristics over recommended operating free-air temperature range (unless otherwise noted)

PARAMETER	TEST CONDITIONS[†]			SN5405			SN7405			UNIT
				MIN	TYP[‡]	MAX	MIN	TYP[‡]	MAX	
V_{IK}	V_{CC} = MIN,	I_I = −12 mA				−1.5			−1.5	V
I_{OH}	V_{CC} = MIN,	V_{IL} = 0.8 V,	V_{OH} = 5.5 V						0.25	mA
	V_{CC} = MIN,	V_{IL} = 0.7 V,	V_{OH} = 5.5 V			0.25				
V_{OL}	V_{CC} = MIN,	V_{IH} = 2 V,	I_{OL} = 16 mA		0.2	0.4		0.2	0.4	V
I_I	V_{CC} = MAX,	V_I = 5.5 V				1			1	mA
I_{IH}	V_{CC} = MAX,	V_I = 2.4 V				40			40	μA
I_{IL}	V_{CC} = MAX,	V_I = 0.4 V				−1.6			−1.6	mA
I_{CCH}	V_{CC} = MAX,	V_I = 0			6	12		6	12	mA
I_{CCL}	V_{CC} = MAX,	V_I = 4.5 V			18	33		18	33	mA

[†] For conditions shown as MIN or MAX, use the appropriate value specified under recommended operating conditions.
[‡] All typical values are at V_{CC} = 5 V, T_A = 25°C

switching characteristics, V_{CC} = 5 V, T_A = 25°C (see note 2)

PARAMETER	FROM (INPUT)	TO (OUTPUT)	TEST CONDITIONS		MIN	TYP	MAX	UNIT
t_{PLH}	A	Y	R_L = 4 kΩ,	C_L = 15 pF		40	55	ns
t_{PHL}			R_L = 400 Ω,	C_L = 15 pF		8	15	ns

NOTE 2: Load circuits and voltage waveforms are shown in Section 1.

TTL Devices

TEXAS
INSTRUMENTS
POST OFFICE BOX 655012 • DALLAS, TEXAS 75265

SN5408, SN7408
QUADRUPLE 2-INPUT POSITIVE-AND GATES

recommended operating conditions

		SN5408			SN7408			UNIT
		MIN	NOM	MAX	MIN	NOM	MAX	
V_{CC}	Supply voltage	4.5	5	5.5	4.75	5	5.25	V
V_{IH}	High-level input voltage	2			2			V
V_{IL}	Low-level input voltage			0.8			0.8	V
I_{OH}	High-level output current			−0.8			−0.8	mA
I_{OL}	Low-level output current			16			16	mA
T_A	Operating free-air temperature	−55		125	0		70	°C

electrical characteristics over recommended operating free-air temperature range (unless otherwise noted)

PARAMETER	TEST CONDITIONS†			SN5408			SN7408			UNIT
			MIN	TYP‡	MAX	MIN	TYP‡	MAX	
V_{IK}	V_{CC} = MIN,	I_I = −12 mA			−1.5			−1.5	V
V_{OH}	V_{CC} = MIN,	V_{IH} = 2 V, I_{OH} = −0.8 mA	2.4	3.4		2.4	3.4		V
V_{OL}	V_{CC} = MIN,	V_{IL} = 0.8 V, I_{OL} = 16 mA		0.2	0.4		0.2	0.4	V
I_I	V_{CC} = MAX,	V_I = 5.5 V			1			1	mA
I_{IH}	V_{CC} = MAX,	V_I = 2.4 V			40			40	μA
I_{IL}	V_{CC} = MAX,	V_I = 0.4 V			−1.6			−1.6	mA
I_{OS}§	V_{CC} = MAX		−20		−55	−18		−55	mA
I_{CCH}	V_{CC} = MAX,	V_I = 4.5 V		11	21		11	21	mA
I_{CCL}	V_{CC} = MAX,	V_I = 0 V		20	33		20	33	mA

† For conditions shown as MIN or MAX, use the appropriate value specified under recommended operating conditions.
‡ All typical values are at V_{CC} = 5 V, T_A = 25°C.
§ Not more than one output should be shorted at a time.

switching characteristics, V_{CC} = 5 V, T_A = 25°C (see note 2)

PARAMETER	FROM (INPUT)	TO (OUTPUT)	TEST CONDITIONS		MIN	TYP	MAX	UNIT
t_{PLH}	A or B	Y	R_L = 400 Ω,	C_L = 15 pF		17.5	27	ns
t_{PHL}						12	19	ns

NOTE 2: Load circuits and voltage waveforms are shown in Section 1.

TTL Devices

TEXAS
INSTRUMENTS
POST OFFICE BOX 655012 • DALLAS, TEXAS 75265

SN5432, SN7432
QUADRUPLE 2-INPUT POSITIVE-OR GATES

recommended operating conditions

		SN5432			SN7432			UNIT
		MIN	NOM	MAX	MIN	NOM	MAX	
V_{CC}	Supply voltage	4.5	5	5.5	4.75	5	5.25	V
V_{IH}	Hgh-level input voltage	2			2			V
V_{IL}	Low-level imput voltage			0.8			0.8	V
I_{OH}	High-level output current			-0.8			-0.8	mA
I_{OL}	Low-level output current			16			16	mA
T_A	Operating free-air temperature	-55		125	0		70	°C

electrical characteristics over recommended operating free-air temperature range (unless otherwise noted)

PARAMETER	TEST CONDITIONS†			SN5432			SN7432			UNIT
				MIN	TYP‡	MAX	MIN	TYP‡	MAX	
V_{IK}	V_{CC} = MIN,	$I_I = -12$ mA				-1.5			-1.5	V
V_{OH}	V_{CC} = MIN,	$V_{IH} = 2$ V,	$I_{OH} = -0.8$ mA	2.4	3.4		2.4	3.4		V
V_{OL}	V_{CC} = MIN,	$V_{IL} = 0.8$ V,	$I_{OL} = 16$ mA		0.2	0.4		0.2	0.4	V
I_I	V_{CC} = MAX,	$V_I = 5.5$ V				1			1	mA
I_{IH}	V_{CC} = MAX,	$V_I = 2.4$ V				40			40	μA
I_{IL}	V_{CC} = MAX,	$V_I = 0.4$ V				-1.6			-1.6	mA
I_{OS}§	V_{CC} = MAX			-20		-55	-18		-55	mA
I_{CCH}	V_{CC} = MAX,	See Note 2			15	22		15	22	mA
I_{CCL}	V_{CC} = MAX,	$V_I = 0$ V			23	38		23	38	mA

† For conditions shown as MIN or MAX, use the appropriate value specified under recommended operating conditions.
‡ All typical values are at $V_{CC} = 5$ V, $T_A = 25°C$.
§ Not more than one output should be shorted at a time.
NOTE 2: One input at 4.5 V, all others at GND.

switching characteristics, $V_{CC} = 5$ V, $T_A = 25°C$ (see note 3)

PARAMETER	FROM (INPUT)	TO (OUTPUT)	TEST CONDITIONS		MIN	TYP	MAX	UNIT
t_{PLH}	A or B	Y	$R_L = 400$ Ω,	$C_L = 15$ pF		10	15	ns
t_{PHL}						14	22	ns

NOTE 3: Load circuits and voltage waveforms are shown in Section 1.

TEXAS
INSTRUMENTS
POST OFFICE BOX 655012 • DALLAS, TEXAS 75265

TTL Devices

SN5474, SN7474
DUAL D-TYPE POSITIVE-EDGE-TRIGGERED FLIP-FLOPS WITH PRESET AND CLEAR

recommended operating conditions

		SN5474			SN7474			UNIT
		MIN	NOM	MAX	MIN	NOM	MAX	
V_{CC}	Supply voltage	4.5	5	5.5	4.75	5	5.25	V
V_{IH}	High-level input voltage	2			2			V
V_{IL}	Low-level input voltage			0.8			0.8	V
I_{OH}	High-level output current			−0.4			−0.4	mA
I_{OL}	Low-level output current			16			16	mA
t_w	Pulse duration CLK high	30			30			ns
	Pulse duration CLK low	37			37			
	Pulse duration $\overline{PRE}$ or $\overline{CLR}$ low	30			30			
t_{su}	Input setup time before CLK ↑	20			20			ns
t_h	Input hold time-data after CLK ↑	5			5			ns
T_A	Operating free-air temperature	−55		125	0		70	°C

electrical characteristics over recommended operating free-air temperature range (unless otherwise noted)

PARAMETER		TEST CONDITIONS[†]			SN5474			SN7474			UNIT
					MIN	TYP[‡]	MAX	MIN	TYP[‡]	MAX	
V_{IK}		V_{CC} = MIN,	I_I = −12 mA				−1.5			−1.5	V
V_{OH}		V_{CC} = MIN,	V_{IH} = 2 V,	V_{IL} = 0.8 V, I_{OH} = −0.4 mA	2.4	3.4		2.4	3.4		V
V_{OL}		V_{CC} = MIN,	V_{IH} = 2 V,	V_{IL} = 0.8 V, I_{OL} = 16 mA		0.2	0.4		0.2	0.4	V
I_I		V_{CC} = MAX,	V_I = 5.5 V				1			1	mA
I_{IH}	D						40			40	μA
	$\overline{CLR}$	V_{CC} = MAX,	V_I = 2.4 V				120			120	
	All Other						80			80	
I_{IL}	D						−1.6			−1.6	mA
	$\overline{PRE}$[§]	V_{CC} = MAX,	V_I = 0.4 V				−1.6			−1.6	
	$\overline{CLR}$[§]						−3.2			−3.2	
	CLK						−3.2			−3.2	
I_{OS}[¶]		V_{CC} = MAX			−20		−57	−18		−57	mA
I_{CC}[#]		V_{CC} = MAX,	See Note 2			8.5	15		8.5	15	mA

[†]For conditions shown as MIN or MAX, use the appropriate value specified under recommended operating conditions.
[‡]All typical values are at V_{CC} = 5 V, T_A = 25 °C.
[§]Clear is tested with preset high and preset is tested with clear high.
[¶]Not more than one output should be shown at a time.
[#]Average per flip-flop.
NOTE 2: With all outputs open, I_{CC} is measured with the Q and $\overline{Q}$ outputs high in turn. At the time of measurement, the clock input is grounded.

switching characteristics, V_{CC} = 5 V, T_A = 25°C (see note 3)

PARAMETER	FROM (INPUT)	TO (OUTPUT)	TEST CONDITIONS		MIN	TYP	MAX	UNIT
f_{max}					15	25		MHz
t_{PLH}	$\overline{PRE}$ or $\overline{CLR}$	Q or $\overline{Q}$	R_L = 400 Ω,	C_L = 15 pF			25	ns
t_{PHL}							40	ns
t_{PLH}	CLK	Q or $\overline{Q}$				14	25	ns
t_{PHL}						20	40	ns

NOTE 3: Load circuits and voltage waveforms are shown in Section 1.

TEXAS INSTRUMENTS
POST OFFICE BOX 655012 • DALLAS, TEXAS 75265

TTL Devices (vertical left margin)

SN54109, SN74109
DUAL J-K̄ POSITIVE-EDGE-TRIGGERED FLIP-FLOPS WITH PRESET AND CLEAR

TTL Devices

recommended operating conditions

			SN54109			SN74109			UNIT
			MIN	NOM	MAX	MIN	NOM	MAX	
V_{CC}	Supply voltage		4.5	5	5.5	4.75	5	5.25	V
V_{IH}	High-level input voltage		2			2			V
V_{IL}	Low-level input voltage				0.8			0.8	V
I_{OH}	High-level output current				−0.8			−0.8	mA
I_{OL}	Low-level output current				16			16	mA
t_W	Pulse duration	CLK high or low	20			20			ns
		$\overline{PRE}$ or $\overline{CLR}$ low	20			20			
t_{su}	Input setup time before CLK↑		10			10			ns
t_h	Input hold time-data after CLK↑		6			6			ns
T_A	Operating free-air temperature		−55		125	0		70	°C

electrical characteristics over recommended operating free-air temperature range (unless otherwise noted)

PARAMETER		TEST CONDITIONS†			SN54109			SN74109			UNIT
					MIN	TYP‡	MAX	MIN	TYP‡	MAX	
V_{IK}		V_{CC} = MIN,	I_I = −12 mA				−1.5			−1.5	V
V_{OH}		V_{CC} = MIN, I_{OH} = −0.8 mA	V_{IH} = 2 V,	V_{IL} = 0.8 V,	2.4	3.4		2.4	3.4		V
V_{OL}		V_{CC} = MIN, I_{OL} = 16 mA	V_{IH} = 2 V,	V_{IL} = 0.8 V,		0.2	0.4		0.2	0.4	V
I_I		V_{CC} = MAX,	V_I = 5.5 V				1			1	mA
I_{IH}	J or $\overline{K}$	V_{CC} = MAX,	V_I = 2.4 V				40			40	μA
	$\overline{CLR}$						160			160	
	$\overline{PRE}$ or CLK						80			80	
I_{IL}	J or $\overline{K}$	V_{CC} = MAX,	V_I = 0.4 V				−1.6			−1.6	mA
	$\overline{CLR}$¶						−4.8			−4.8	
	$\overline{PRE}$¶						−3.2			−3.2	
	CLK						−3.2			−3.2	
I_{OS}§		V_{CC} = MAX			−30		−85	−30		−85	mA
I_{CC}#		V_{CC} = MAX,	See Note 2			9	15		9	15	mA

† For conditions shown as MIN or MAX, use the appropriate value specified under recommended operating conditions.
‡ All typical values are at V_{CC} = 5 V, T_A = 25°C.
§ Not more than one output should be shorted at a time.
¶ Clear is tested with preset high and preset is tested with clear high.
# Average per flip-flop.
NOTE 2: With all outputs open, I_{CC} is measured with the Q and $\overline{Q}$ outputs high in turn. At the time of measurement, the clock input is grounded.

switching characteristics, V_{CC} = 5 V, T_A = 25°C (see note 3)

PARAMETER	FROM (INPUT)	TO (OUTPUT)	TEST CONDITIONS		MIN	TYP	MAX	UNIT
f_{max}					25	33		MHz
t_{PLH}	$\overline{PRE}$	Q				10	15	ns
t_{PHL}		$\overline{Q}$				23	35	ns
t_{PLH}	$\overline{CLR}$	$\overline{Q}$	R_L = 400 Ω,	C_L = 15 pF		10	15	ns
t_{PHL}		Q				17	25	ns
t_{PLH}	CLK	Q or $\overline{Q}$				10	16	ns
t_{PHL}						18	28	ns

NOTE 3: Load circuits and voltage waveforms are shown in Section 1.

TEXAS INSTRUMENTS
POST OFFICE BOX 655012 • DALLAS, TEXAS 75265

TYPES SN54LS00, SN74LS00
QUADRUPLE 2-INPUT POSITIVE-NAND GATES

recommended operating conditions

		SN54LS00			SN74LS00			UNIT
		MIN	NOM	MAX	MIN	NOM	MAX	
V_{CC}	Supply voltage	4.5	5	5.5	4.75	5	5.25	V
V_{IH}	High-level input voltage	2			2			V
V_{IL}	Low-level input voltage			0.7			0.8	V
I_{OH}	High-level output current			−0.4			−0.4	mA
I_{OL}	Low-level output current			4			8	mA
T_A	Operating free-air temperature	−55		125	0		70	°C

electrical characteristics over recommended operating free-air temperature range (unless otherwise noted)

PARAMETER	TEST CONDITIONS †			SN54LS00			SN74LS00			UNIT
			MIN	TYP‡	MAX	MIN	TYP‡	MAX	
V_{IK}	V_{CC} = MIN,	I_I = −18 mA				−1.5			−1.5	V
V_{OH}	V_{CC} = MIN,	V_{IL} = MAX, I_{OH} = −0.4 mA		2.5	3.4		2.7	3.4		V
V_{OL}	V_{CC} = MIN,	V_{IH} = 2 V, I_{OL} = 4 mA			0.25	0.4		0.25	0.4	V
	V_{CC} = MIN,	V_{IH} = 2 V, I_{OL} = 8 mA						0.35	0.5	
I_I	V_{CC} = MAX,	V_I = 7 V				0.1			0.1	mA
I_{IH}	V_{CC} = MAX,	V_I = 2.7 V				20			20	μA
I_{IL}	V_{CC} = MAX,	V_I = 0.4 V				−0.4			−0.4	mA
I_{OS} §	V_{CC} = MAX			−20		−100	−20		−100	mA
I_{CCH}	V_{CC} = MAX,	V_I = 0 V			0.8	1.6		0.8	1.6	mA
I_{CCL}	V_{CC} = MAX,	V_I = 4.5 V			2.4	4.4		2.4	4.4	mA

† For conditions shown as MIN or MAX, use the appropriate value specified under recommended operating conditions.
‡ All typical values are at V_{CC} = 5 V, T_A = 25°C
§ Not more than one output should be shorted at a time, and the duration of the short-circuit should not exceed one second.

switching characteristics, V_{CC} = 5 V, T_A = 25°C (see note 2)

PARAMETER	FROM (INPUT)	TO (OUTPUT)	TEST CONDITIONS		MIN	TYP	MAX	UNIT
t_{PLH}	A or B	Y	R_L = 2 kΩ,	C_L = 15 pF		9	15	ns
t_{PHL}						10	15	ns

NOTE 2: See General Information Section for load circuits and voltage waveforms.

TEXAS INSTRUMENTS
POST OFFICE BOX 225012 • DALLAS, TEXAS 75265

TTL DEVICES

TYPES SN54LS02, SN74LS02
QUADRUPLE 2-INPUT POSITIVE-NOR GATES

recommended operating conditions

		SN54LS02			SN74LS02			UNIT
		MIN	NOM	MAX	MIN	NOM	MAX	
V_{CC}	Supply voltage	4.5	5	5.5	4.75	5	5.25	V
V_{IH}	High-level input voltage	2			2			V
V_{IL}	Low-level input voltage			0.7			0.8	V
I_{OH}	High-level output current			-0.4			-0.4	mA
I_{OL}	Low-level output current			4			8	mA
T_A	Operating free-air temperature	-55		125	0		70	°C

electrical characteristics over recommended operating free-air temperature range (unless otherwise noted)

PARAMETER	TEST CONDITIONS †			SN54LS02			SN74LS02			UNIT
				MIN	TYP‡	MAX	MIN	TYP‡	MAX	
V_{IK}	V_{CC} = MIN,	I_I = -18 mA				-1.5			-1.5	V
V_{OH}	V_{CC} = MIN,	V_{IL} = MAX,	I_{OH} = -0.4 mA	2.5	3.4		2.7	3.4		V
V_{OL}	V_{CC} = MIN,	V_{IH} = 2 V,	I_{OL} = 4 mA		0.25	0.4		0.25	0.4	V
	V_{CC} = MIN,	V_{IH} = 2 V,	I_{OL} = 8 mA					0.35	0.5	
I_I	V_{CC} = MAX,	V_I = 7 V				0.1			0.1	mA
I_{IH}	V_{CC} = MAX,	V_I = 2.7 V				20			20	µA
I_{IL}	V_{CC} = MAX,	V_I = 0.4 V				-0.4			-0.4	mA
I_{OS} §	V_{CC} = MAX			-20		-100	-20		-100	mA
I_{CCH}	V_{CC} = MAX,	V_I = 0 V			1.6	3.2		1.6	3.2	mA
I_{CCL}	V_{CC} = MAX,	See Note 2			2.8	5.4		2.8	5.4	mA

† For conditions shown as MIN or MAX, use the appropriate value specified under recommended operating conditions.
‡ All typical values are at V_{CC} = 5 V, T_A = 25°C
§ Not more than one output should be shorted at a time, and the duration of the short-circuit should not exceed one second.
NOTE 2: One input at 4.5 V, all others at GND.

switching characteristics, V_{CC} = 5 V, T_A = 25°C (see note 3)

PARAMETER	FROM (INPUT)	TO (OUTPUT)	TEST CONDITIONS		MIN	TYP	MAX	UNIT
t_{PLH}	A or B	Y	R_L = 2 kΩ,	C_L = 15 pF		10	15	ns
t_{PHL}						10	15	ns

NOTE 3: See General Information Section for load circuits and voltage waveforms.

TTL DEVICES

TEXAS INSTRUMENTS
POST OFFICE BOX 225012 • DALLAS, TEXAS 75265

TYPES SN54LS04, SN74LS04
HEX INVERTERS

recommended operating conditions

		SN54LS04			SN74LS04			UNIT
		MIN	NOM	MAX	MIN	NOM	MAX	
V_{CC}	Supply voltage	4.5	5	5.5	4.75	5	5.25	V
V_{IH}	High-level input voltage	2			2			V
V_{IL}	Low-level input voltage			0.7			0.8	V
I_{OH}	High-level output current			−0.4			−0.4	mA
I_{OL}	Low-level output current			4			8	mA
T_A	Operating free-air temperature	−55		125	0		70	°C

electrical characteristics over recommended operating free-air temperature range (unless otherwise noted)

PARAMETER	TEST CONDITIONS †			SN54LS04			SN74LS04			UNIT
			MIN	TYP ‡	MAX	MIN	TYP ‡	MAX	
V_{IK}	V_{CC} = MIN,	I_I = −18 mA			−1.5			−1.5	V
V_{OH}	V_{CC} = MIN, V_{IL} = MAX,	I_{OH} = −0.4 mA	2.5	3.4		2.7	3.4		V
V_{OL}	V_{CC} = MIN, V_{IH} = 2 V,	I_{OL} = 4 mA		0.25	0.4			0.4	V
	V_{CC} = MIN, V_{IH} = 2 V,	I_{OL} = 8 mA					0.25	0.5	
I_I	V_{CC} = MAX,	V_I = 7 V			0.1			0.1	mA
I_{IH}	V_{CC} = MAX,	V_I = 2.7 V			20			20	μA
I_{IL}	V_{CC} = MAX,	V_I = 0.4 V			−0.4			−0.4	mA
I_{OS} §	V_{CC} = MAX		−20		−100	−20		−100	mA
I_{CCH}	V_{CC} = MAX,	V_I = 0 V		1.2	2.4		1.2	2.4	mA
I_{CCL}	V_{CC} = MAX,	V_I = 4.5 V		3.6	6.6		3.6	6.6	mA

† For conditions shown as MIN or MAX, use the appropriate value specified under recommended operating conditions.
‡ All typical values are at V_{CC} = 5 V, T_A = 25°C.
§ Not more than one output should be shorted at a time, and the duration of the short-circuit should not exceed one second.

switching characteristics, V_{CC} = 5 V, T_A = 25°C (see note 2)

PARAMETER	FROM (INPUT)	TO (OUTPUT)	TEST CONDITIONS		MIN	TYP	MAX	UNIT
t_{PLH}	A	Y	R_L = 2 kΩ,	C_L = 15 pF		9	15	ns
t_{PHL}						10	15	ns

NOTE 2: See General Information Section for load circuits and voltage waveforms.

TEXAS INSTRUMENTS
POST OFFICE BOX 225012 • DALLAS, TEXAS 75265

Reprinted by permission of Texas Instruments.

TTL DEVICES

TYPES SN54LS05, SN74LS05
HEX INVERTERS WITH OPEN-COLLECTOR OUTPUTS

recommended operating conditions

		SN54LS05			SN74LS05			UNIT
		MIN	NOM	MAX	MIN	NOM	MAX	
V_{CC}	Supply voltage	4.5	5	5.5	4.75	5	5.25	V
V_{IH}	High-level input voltage	2			2			V
V_{IL}	Low-level input voltage			0.7			0.8	V
V_{OH}	High-level output voltage			5.5			5.5	V
I_{OL}	Low-level output current			4			8	mA
T_A	Operating free-air temperature	-55		125	0		70	$^\circ$C

electrical characteristics over recommended operating free-air temperature range (unless otherwise noted)

PARAMETER	TEST CONDITIONS†			SN54LS05			SN74LS05			UNIT
			MIN	TYP‡	MAX	MIN	TYP‡	MAX		
V_{IK}	V_{CC} = MIN,	$I_I = -18$ mA			-1.5			-1.5		V
I_{OH}	V_{CC} = MIN,	V_{IL} = MAX, V_{OH} = 5.5 V			0.1			0.1		mA
V_{OL}	V_{CC} = MIN,	V_{IH} = 2 V, I_{OL} = 4 mA		0.25	0.4		0.25	0.4		V
	V_{CC} = MIN,	V_{IH} = 2 V, I_{OL} = 8 mA					0.35	0.5		
I_I	V_{CC} = MAX,	V_I = 7 V			0.1			0.1		mA
I_{IH}	V_{CC} = MAX,	V_I = 2.7 V			20			20		μA
I_{IL}	V_{CC} = MAX,	V_I = 0.4 V			-0.4			-0.4		mA
I_{CCH}	V_{CC} = MAX,	V_I = 0 V		1.2	2.4		1.2	2.4		mA
I_{CCL}	V_{CC} = MAX,	V_I = 4.5 V		3.6	6.6		3.6	6.6		mA

† For conditions shown as MIN or MAX, use the appropriate value specified under recommended operating conditions.
‡ All typical values are at V_{CC} = 5 V, T_A = 25°C.

switching characteristics, V_{CC} = 5 V, T_A = 25°C (see note 2)

PARAMETER	FROM (INPUT)	TO (OUTPUT)	TEST CONDITIONS		MIN	TYP	MAX	UNIT
t_{PLH}	A	Y	R_L = 2 kΩ,	C_L = 15 pF		17	32	ns
t_{PHL}						15	28	ns

NOTE 2: See General Information Section for load circuits and voltage waveforms.

TEXAS
INSTRUMENTS
POST OFFICE BOX 225012 ● DALLAS, TEXAS 75265

TTL DEVICES

Reprinted by permission of Texas Instruments.

TYPES SN54LS08, SN74LS08
QUADRUPLE 2-INPUT POSITIVE-AND GATES

recommended operating conditions

		SN54LS08			SN74LS08			UNIT
		MIN	NOM	MAX	MIN	NOM	MAX	
V_{CC}	Supply voltage	4.5	5	5.5	4.75	5	5.25	V
V_{IH}	High-level input voltage	2			2			V
V_{IL}	Low-level input voltage			0.7			0.8	V
I_{OH}	High-level output current			−0.4			−0.4	mA
I_{OL}	Low-level output current			4			8	mA
T_A	Operating free-air temperature	−55		125	0		70	°C

electrical characteristics over recommended operating free-air temperature range (unless otherwise noted)

PARAMETER	TEST CONDITIONS †			SN54LS08			SN74LS08			UNIT
				MIN	TYP‡	MAX	MIN	TYP‡	MAX	
V_{IK}	V_{CC} = MIN,	I_I = −18 mA				−1.5			−1.5	V
V_{OH}	V_{CC} = MIN,	V_{IH} = 2 V,	I_{OH} = −0.4 mA	2.5	3.4		2.7	3.4		V
V_{OL}	V_{CC} = MIN,	V_{IL} = MAX,	I_{OL} = 4 mA		0.25	0.4		0.25	0.4	V
	V_{CC} = MIN,	V_{IL} = MAX,	I_{OL} = 8 mA					0.35	0.5	
I_I	V_{CC} = MAX,	V_I = 7 V				0.1			0.1	mA
I_{IH}	V_{CC} = MAX,	V_I = 2.7 V				20			20	µA
I_{IL}	V_{CC} = MAX,	V_I = 0.4 V				−0.4			−0.4	mA
I_{OS}§	V_{CC} = MAX			−20		−100	−20		−100	mA
I_{CCH}	V_{CC} = MAX,	V_I = 4.5 V			2.4	4.8		2.4	4.8	mA
I_{CCL}	V_{CC} = MAX,	V_I = 0 V			4.4	8.8		4.4	8.8	mA

† For conditions shown as MIN or MAX, use the appropriate value specified under recommended operating conditions.
‡ All typical values are at V_{CC} = 5 V, T_A = 25°C
§ Not more than one output should be shorted at a time, and the duration of the short-circuit should not exceed one second.

switching characteristics, V_{CC} = 5 V, T_A = 25°C (see note 2)

PARAMETER	FROM (INPUT)	TO (OUTPUT)	TEST CONDITIONS		MIN	TYP	MAX	UNIT
t_{PLH}	A or B	Y	R_L = 2 kΩ,	C_L = 15 pF		8	15	ns
t_{PHL}						10	20	ns

NOTE 2: See General Information Section for load circuits and voltage waveforms.

TEXAS
INSTRUMENTS
POST OFFICE BOX 225012 • DALLAS, TEXAS 75265

TYPES SN54LS32, SN74LS32
QUADRUPLE 2-INPUT POSITIVE-OR GATES

recommended operating conditions

		SN54LS32			SN74LS32			UNIT
		MIN	NOM	MAX	MIN	NOM	MAX	
V_{CC}	Supply voltage	4.5	5	5.5	4.75	5	5.25	V
V_{IH}	Hgh-level input voltage	2			2			V
V_{IL}	Low-level input voltage			0.7			0.8	V
I_{OH}	High-level output current			-0.4			-0.4	mA
I_{OL}	Low-level output current			4			8	mA
T_A	Operating free-air temperature	-55		125	0		70	°C

electrical characteristics over recommended operating free-air temperature range (unless otherwise noted)

PARAMETER	TEST CONDITIONS†			SN54LS32			SN74LS32			UNIT
			MIN	TYP‡	MAX	MIN	TYP‡	MAX	
V_{IK}	V_{CC} = MIN,	$I_I = -18$ mA				-1.5			-1.5	V
V_{OH}	V_{CC} = MIN,	V_{IH} = 2 V,	$I_{OH} = -0.4$ mA	2.5	3.4		2.7	3.4		V
V_{OL}	V_{CC} = MIN,	V_{IL} = MAX,	I_{OL} = 4 mA		0.25	0.4		0.25	0.4	V
	V_{CC} = MIN,	V_{IL} = MAX,	I_{OL} = 8 mA					0.35	0.5	
I_I	V_{CC} = MAX,	V_I = 7 V				0.1			0.1	mA
I_{IH}	V_{CC} = MAX,	V_I = 2.7 V				20			20	μA
I_{IL}	V_{CC} = MAX,	V_I = 0.4 V				-0.4			-0.4	mA
I_{OS} §	V_{CC} = MAX			-20		-100	-20		-100	mA
I_{CCH}	V_{CC} = MAX,	See Note 2			3.1	6.2		3.1	6.2	mA
I_{CCL}	V_{CC} = MAX,	V_I = 0 V			4.9	9.8		4.9	9.8	mA

† For conditions shown as MIN or MAX, use the appropriate value specified under recommended operating conditions.
‡ All typical values are at V_{CC} = 5 V, $T_A = 25°C$.
§ Not more than one output should be shorted at a time and the duration of the short-circuit should not exceed one second.
NOTE 2: One input at 4.5 V, all others at GND.

switching characteristics, V_{CC} = 5 V, T_A = 25°C (see note 3)

PARAMETER	FROM (INPUT)	TO (OUTPUT)	TEST CONDITIONS		MIN	TYP	MAX	UNIT
t_{PLH}	A or B	Y	$R_L = 2$ kΩ,	$C_L = 15$ pF		14	22	ns
t_{PHL}						14	22	ns

NOTE 3: See General Information Section for load circuits and voltage waveforms.

TEXAS INSTRUMENTS
POST OFFICE BOX 225012 ● DALLAS, TEXAS 75265

SN54LS74A, SN74LS74A
DUAL D-TYPE POSITIVE-EDGE-TRIGGERED FLIP-FLOPS WITH PRESET AND CLEAR

recommended operating conditions

			SN54LS74A			SN74LS74A			UNIT
			MIN	NOM	MAX	MIN	NOM	MAX	
V_{CC}	Supply voltage		4.5	5	5.5	4.75	5	5.25	V
V_{IH}	High-level input voltage		2			2			V
V_{IL}	Low-level input voltage				0.7			0.8	V
I_{OH}	High-level output current				−0.4			−0.4	mA
I_{OL}	Low-level output current				4			8	mA
f_{clock}	Clock frequency		0		25	0		25	MHz
t_w	Pulse duration	CLK high	25			25			ns
		$\overline{PRE}$ or $\overline{CLR}$ low	25			25			
t_{su}	Setup time-before CLK↑	High-level data	20			20			ns
		Low-level data	20			20			
t_h	Hold time-data after CLK↑		5			5			ns
T_A	Operating free-air temperature		−55		125	0		70	°C

electrical characteristics over recommended operating free-air temperature range (unless otherwise noted)

PARAMETER		TEST CONDITIONS[†]			SN54LS74A			SN74LS74A			UNIT
				MIN	TYP[‡]	MAX	MIN	TYP[‡]	MAX		
V_{IK}		V_{CC} = MIN,	I_I = −18 mA				−1.5			−1.5	V
V_{OH}		V_{CC} = MIN, V_{IH} = 2 V, V_{IL} = MAX, I_{OH} = −0.4 mA			2.5	3.4		2.7	3.4		V
V_{OL}		V_{CC} = MIN, V_{IL} = MAX, V_{IH} = 2 V, I_{OL} = 4 mA				0.25	0.4		0.25	0.4	V
		V_{CC} = MIN, V_{IL} = MAX, V_{IH} = 2 V, I_{OL} = 8 mA							0.35	0.5	
I_I	D or CLK	V_{CC} = MAX,	V_I = 7 V				0.1			0.1	mA
	$\overline{CLR}$ or $\overline{PRE}$						0.2			0.2	
I_{IH}	D or CLK	V_{CC} = MAX,	V_I = 2.7 V				20			20	µA
	$\overline{CLR}$ or $\overline{PRE}$						40			40	
I_{IL}	D or CLK	V_{CC} = MAX,	V_I = 0.4 V				−0.4			−0.4	mA
	$\overline{CLR}$ or $\overline{PRE}$						−0.8			−0.8	
I_{OS}§		V_{CC} = MAX,	See Note 4		−20		−100	−20		−100	mA
I_{CC} (Total)		V_{CC} = MAX,	See Note 2			4	8		4	8	mA

† For conditions shown as MIN or MAX, use the appropriate value specified under recommended operating conditions.
‡ All typical values are at V_{CC} = 5 V, T_A = 25°C.
§ Not more than one output should be shorted at a time, and the duration of the short circuit should not exceed one second.
NOTE 2: With all outputs open, I_{CC} is measured with the Q and $\overline{Q}$ outputs high in turn. At the time of measurement, the clock input is grounded.
NOTE 4: For certain devices where state commutation can be caused by shorting an output to ground, an equivalent test may be performed with V_O = 2.25 V and 2.125 V for the 54 family and the 74 family, respectively, with the minimum and maximum limits reduced to one half of their stated values.

switching characteristics, V_{CC} = 5 V, T_A = 25°C (see note 3)

PARAMETER	FROM (INPUT)	TO (OUTPUT)	TEST CONDITIONS		MIN	TYP	MAX	UNIT
f_{max}					25	33		MHz
t_{PLH}	$\overline{CLR}$, $\overline{PRE}$ or CLK	Q or $\overline{Q}$	R_L = 2 kΩ,	C_L = 15 pF		13	25	ns
t_{PHL}						25	40	ns

Note 3: Load circuits and voltage waveforms are shown in Section 1.

TTL Devices (side text)

TEXAS INSTRUMENTS
POST OFFICE BOX 655012 • DALLAS, TEXAS 75265

SN54LS109A, SN74LS109A
DUAL J-$\overline{\text{K}}$ POSITIVE-EDGE-TRIGGERED FLIP-FLOPS WITH PRESET AND CLEAR

recommended operating conditions

			SN54LS109A			SN74LS109A			UNIT
			MIN	NOM	MAX	MIN	NOM	MAX	
V_{CC}	Supply voltage		4.5	5	5.5	4.75	5	5.25	V
V_{IH}	High-level input voltage		2			2			V
V_{IL}	Low-level input voltage				0.7			0.8	V
I_{OH}	High-level output current				−0.4			−0.4	mA
I_{OL}	Low-level output current				4			8	mA
f_{clock}	Clock frequency		0		25	0		25	MHz
t_w	Pulse duration	CLK high	25			25			ns
		$\overline{\text{PRE}}$ or $\overline{\text{CLR}}$ low	25			25			
t_{su}	Setup time before CLK ↑	High-level data	35			35			ns
		Low-level data	25			25			
t_h	Hold time-data after CLK ↑		5			5			ns
T_A	Operating free-air temperature		−55		125	0		70	°C

electrical characteristics over recommended operating free-air temperature range (unless otherwise noted)

PARAMETER		TEST CONDITIONS[†]			SN54LS109A			SN74LS109A			UNIT
					MIN	TYP[‡]	MAX	MIN	TYP[‡]	MAX	
V_{IK}		V_{CC} = MIN,	I_I = −18 mA				−1.5			−1.5	V
V_{OH}		V_{CC} = MIN,	V_{IH} = 2 V,	V_{IL} = MAX,	2.5	3.4		2.7	3.4		V
		I_{OH} = −0.4 mA									
V_{OL}		V_{CC} = MIN,	V_{IL} = MAX,	V_{IH} = 2 V,		0.25	0.4		0.25	0.4	V
		I_{OL} = 4 mA									
		V_{CC} = MIN,	V_{IL} = MAX,	V_{IH} = 2 V,					0.35	0.5	
		I_{OL} = 8 mA									
I_I	J, $\overline{\text{K}}$ or CLK	V_{CC} = MAX,	V_I = 7 V				0.1			0.1	mA
	$\overline{\text{CLR}}$ or $\overline{\text{PRE}}$						0.2			0.2	
I_{IH}	J, $\overline{\text{K}}$ or CLK	V_{CC} = MAX,	V_I = 2.7 V				20			20	μA
	$\overline{\text{CLR}}$ or $\overline{\text{PRE}}$						40			40	
I_{IL}	J, $\overline{\text{K}}$ or CLK	V_{CC} = MAX,	V_I = 0.4 V				−0.4			−0.4	mA
	$\overline{\text{CLR}}$ or $\overline{\text{PRE}}$						−0.8			−0.8	
I_{OS} §		V_{CC} = MAX,	See Note 4		−20		−100	−20		−100	mA
I_{CC} (Total)		V_{CC} = MAX,	See Note 2			4	8		4	8	mA

† For conditions shown as MIN or MAX, use the appropriate value specified under recommended operating conditions.
‡ All typical values are at V_{CC} = 5 V, T_A = 25°C.
§ Not more than one output should be shorted at a time, and the duration of the short circuit should not exceed one second.
NOTE 2: With all outputs open, I_{CC} is measured wtih the Q and $\overline{Q}$ outputs high in turn. At the time of measurement, the clock input is grounded.
NOTE 4: For certain devices where state commutation can be caused by shorting an output to ground, an equivalent test may be performed with V_O = 2.25 V and 2.125 V for the 54 family and the 74 family, respectively with the minimum and maximum limits reduced to one half of their stated values.

switching characteristics, V_{CC} = 5 V, T_A = 25°C (see note 3)

PARAMETER	FROM (INPUT)	TO (OUTPUT)	TEST CONDITIONS		MIN	TYP	MAX	UNIT
f_{max}			R_L = 2 kΩ,	C_L = 15 pF	25	33		MHz
t_{PLH}	$\overline{\text{CLR}}$, $\overline{\text{PRE}}$	Q or $\overline{Q}$				13	25	ns
t_{PHL}	or CLK					25	40	ns

NOTE 3: Load circuits and voltage waveforms are shown in Section 1.

TEXAS
INSTRUMENTS

POST OFFICE BOX 655012 • DALLAS, TEXAS 75265

TTL Devices

SN54AS00, SN74AS00
QUADRUPLE 2-INPUT POSITIVE-NAND GATES

absolute maximum ratings over operating free-air temperature range (unless otherwise noted)

Supply voltage, V_{CC} . . . . 7 V
Input voltage . . . . 7 V
Operating free-air temperature range: SN54AS00 . . . . −55 °C to 125 °C
SN74AS00 . . . . 0 °C to 70 °C
Storage temperature range . . . . −65 °C to 150 °C

recommended operating conditions

		SN54AS00			SN74AS00			UNIT
		MIN	NOM	MAX	MIN	NOM	MAX	
V_{CC}	Supply voltage	4.5	5	5.5	4.5	5	5.5	V
V_{IH}	High-level input voltage	2			2			V
V_{IL}	Low-level input voltage			0.8			0.8	V
I_{OH}	High-level output current			−2			−2	mA
I_{OL}	Low-level output current			20			20	mA
T_A	Operating free-air temperature	−55		125	0		70	°C

electrical characteristics over recommended operating free-air temperature range (unless otherwise noted)

PARAMETER	TEST CONDITIONS		SN54AS00			SN74AS00			UNIT
			MIN	TYP†	MAX	MIN	TYP†	MAX	
V_{IK}	$V_{CC} = 4.5$ V,	$I_I = -18$ mA			−1.2			−1.2	V
V_{OH}	$V_{CC} = 4.5$ V to 5.5 V,	$I_{OH} = -2$ mA	$V_{CC}-2$			$V_{CC}-2$			V
V_{OL}	$V_{CC} = 4.5$ V,	$I_{OL} = 20$ mA		0.35	0.5		0.35	0.5	V
I_I	$V_{CC} = 5.5$ V,	$V_I = 7$ V			0.1			0.1	mA
I_{IH}	$V_{CC} = 5.5$ V,	$V_I = 2.7$ V			20			20	μA
I_{IL}	$V_{CC} = 5.5$ V,	$V_I = 0.4$ V			−0.5			−0.5	mA
I_O‡	$V_{CC} = 5.5$ V,	$V_O = 2.25$ V	−30		−112	−30		−112	mA
I_{CCH}	$V_{CC} = 5.5$ V,	$V_I = 0$ V		2	3.2		2	3.2	mA
I_{CCL}	$V_{CC} = 5.5$ V,	$V_I = 4.5$ V		10.8	17.4		10.8	17.4	mA

†All typical values are at $V_{CC} = 5$ V, $T_A = 25$ °C.
‡The output conditions have been chosen to produce a current that closely approximates one half of the true short-circuit output current, I_{OS}.

switching characteristics (see Note 1)

PARAMETER	FROM (INPUT)	TO (OUTPUT)	$V_{CC} = 4.5$ V to 5.5 V, $C_L = 50$ pF, $R_L = 50$ Ω, $T_A =$ MIN to MAX				UNIT
			SN54AS00		SN74AS00		
			MIN	MAX	MIN	MAX	
t_{PLH}	A or B	Y	1	5	1	4.5	ns
t_{PHL}	A or B	Y	1	5	1	4	

NOTE 1: Load circuit and voltage waveforms are shown in Section 1.

TEXAS INSTRUMENTS
POST OFFICE BOX 655012 • DALLAS, TEXAS 75265

ALS and AS Circuits

SN54AS02, SN74AS02
QUADRUPLE 2-INPUT POSITIVE-NOR GATES

absolute maximum ratings over operating free-air temperature range (unless otherwise noted)

Supply voltage, V_{CC} .... 7 V
Input voltage .... 7 V
Operating free-air temperature range: SN54AS02 .... $-55\,°C$ to $125\,°C$
  SN74AS02 .... $0\,°C$ to $70\,°C$
Storage temperature range .... $-65\,°C$ to $150\,°C$

recommended operating conditions

		SN54AS02			SN74AS02			UNIT
		MIN	NOM	MAX	MIN	NOM	MAX	
V_{CC}	Supply voltage	4.5	5	5.5	4.5	5	5.5	V
V_{IH}	High-level input voltage	2			2			V
V_{IL}	Low-level input voltage			0.8			0.8	V
I_{OH}	High-level output current			-2			-2	mA
I_{OL}	Low-level output current			20			20	mA
T_A	Operating free-air temperature	-55		125	0		70	°C

electrical characteristics over recommended operating-free-air temperature range (unless otherwise noted)

PARAMETER	TEST CONDITIONS		SN54AS02			SN74AS02			UNIT
			MIN	TYP[†]	MAX	MIN	TYP[†]	MAX	
V_{IK}	$V_{CC} = 4.5$ V,	$I_I = -18$ mA			-1.2			-1.2	V
V_{OH}	$V_{CC} = 4.5$ V to 5.5 V,	$I_{OH} = -2$ mA	$V_{CC}-2$			$V_{CC}-2$			V
V_{OL}	$V_{CC} = 4.5$ V,	$I_{OL} = 20$ mA		0.35	0.5		0.35	0.5	V
I_I	$V_{CC} = 5.5$ V,	$V_I = 7$ V			0.1			0.1	mA
I_{IH}	$V_{CC} = 5.5$ V,	$V_I = 2.7$ V			20			20	μA
I_{IL}	$V_{CC} = 5.5$ V,	$V_I = 0.4$ V			-0.5			-0.5	mA
I_O [‡]	$V_{CC} = 5.5$ V,	$V_O = 2.25$ V	-30		-112	-30		-112	mA
I_{CCH}	$V_{CC} = 5.5$ V,	$V_I = 0$ V		3.7	5.9		3.7	5.9	mA
I_{CCL}	$V_{CC} = 5.5$ V,	$V_I = 4.5$ V		12.5	20.1		12.5	20.1	mA

† All typical values are at $V_{CC} = 5$ V, $T_A = 25\,°C$.
‡ The output conditions have been chosen to produce a current that closely approximates one half of the true short-circuit output current, I_{OS}.

switching characteristics (see Note 1)

PARAMETER	FROM (INPUT)	TO (OUTPUT)	$V_{CC} = 4.5$ V to 5.5 V, $C_L = 50$ pF, $R_L = 500\ \Omega$, $T_A = $ MIN to MAX				UNIT
			SN54AS02		SN74AS02		
			MIN	MAX	MIN	MAX	
t_{PLH}	A or B	Y	1	5	1	4.5	ns
t_{PHL}	A or B	Y	1	5	1	4.5	

NOTE 1: Load circuit and voltage waveforms are shown in Section 1.

TEXAS
INSTRUMENTS
POST OFFICE BOX 655012 • DALLAS, TEXAS 75265

Reprinted by permission of Texas Instruments.

ALS and AS Circuits

SN54AS04, SN74AS04
HEX INVERTERS

absolute maximum ratings over operating free-air temperature range (unless otherwise noted)

Supply voltage, V_{CC} . . . . 7 V
Input voltage . . . . 7 V
Operating free-air temperature range: SN54AS04 . . . . $-55\,°C$ to $125\,°C$
                                                          SN74AS04 . . . . $0\,°C$ to $70\,°C$
Storage temperature range . . . . $-65\,°C$ to $150\,°C$

recommended operating conditions

		SN54AS04			SN74AS04			UNIT
		MIN	NOM	MAX	MIN	NOM	MAX	
V_{CC}	Supply voltage	4.5	5	5.5	4.5	5	5.5	V
V_{IH}	High-level input voltage	2			2			V
V_{IL}	Low-level input voltage			0.8			0.8	V
I_{OH}	High-level output current			-2			-2	mA
I_{OL}	Low-level output current			20			20	mA
T_A	Operating free-air temperature	-55		125	0		70	°C

electrical characteristics over recommended operating free-air temperature range (unless otherwise noted)

PARAMETER	TEST CONDITIONS		SN54AS04			SN74AS04			UNIT
			MIN	TYP†	MAX	MIN	TYP†	MAX	
V_{IK}	$V_{CC} = 4.5$ V,	$I_I = -18$ mA			-1.2			-1.2	V
V_{OH}	$V_{CC} = 4.5$ V to 5.5 V,	$I_{OH} = -2$ mA	$V_{CC}-2$			$V_{CC}-2$			V
V_{OL}	$V_{CC} = 4.5$ V,	$I_{OL} = 20$ mA		0.35	0.5		0.35	0.5	V
I_I	$V_{CC} = 5.5$ V,	$V_I = 7$ V			0.1			0.1	mA
I_{IH}	$V_{CC} = 5.5$ V,	$V_I = 2.7$ V			20			20	μA
I_{IL}	$V_{CC} = 5.5$ V,	$V_I = 0.4$ V			-0.5			-0.5	mA
I_O‡	$V_{CC} = 5.5$ V,	$V_O = 2.25$ V	-30		-112	-30		-112	mA
I_{CCH}	$V_{CC} = 5.5$ V,	$V_I = 0$ V		3	4.8		3	4.8	mA
I_{CCL}	$V_{CC} = 5.5$ V,	$V_I = 4.5$ V		14	26.3		14	26.3	mA

†All typical values are at $V_{CC} = 5$ V, $T_A = 25\,°C$.
‡The output conditions have been chosen to produce a current that closely approximates one half of the true short-circuit output current, I_{OS}.

switching characteristics (see Note 1)

PARAMETER	FROM (INPUT)	TO (OUTPUT)	$V_{CC} = 4.5$ V to 5.5 V, $C_L = 50$ pF, $R_L = 500$ Ω, $T_A = $ MIN to MAX				UNIT
			SN54AS04		SN74AS04		
			MIN	MAX	MIN	MAX	
t_{PLH}	A	Y	1	6	1	5	ns
t_{PHL}	A	Y	1	4.5	1	4	ns

NOTE 1. Load circuit and voltage waveforms are shown in Section 1.

TEXAS
INSTRUMENTS
POST OFFICE BOX 655012 • DALLAS, TEXAS 75265

SN54AS08, SN74AS08
QUADRUPLE 2-INPUT POSITIVE-AND GATES

absolute maximum ratings over operating free-air temperature range (unless otherwise noted)

Supply voltage, V_{CC} . . . . 7 V
Input voltage . . . . 7 V
Operating free-air temperature range: SN54AS08 . . . . −55 °C to 125 °C
                         SN74AS08 . . . . 0 °C to 70 °C
Storage temperature range . . . . −65 °C to 150 °C

recommended operating conditions

		SN54AS08			SN74AS08			UNIT
		MIN	NOM	MAX	MIN	NOM	MAX	
V_{CC}	Supply voltage	4.5	5	5.5	4.5	5	5.5	V
V_{IH}	High-level input voltage	2			2			V
V_{IL}	Low-level input voltage			0.8			0.8	mA
I_{OH}	High-level output current.			−2			−2	mA
I_{OL}	Low-level output current			20			20	mA
T_A	Operating free-air temperature	−55		125	0		70	°C

electrical characteristics over recommended operating free-air temperature range (unless otherwise noted)

PARAMETER	TEST CONDITIONS		SN54AS08			SN74AS08			UNIT
			MIN	TYP†	MAX	MIN	TYP†	MAX	
V_{IK}	$V_{CC} = 4.5$ V,	$I_I = -18$ mA			−1.2			−1.2	V
V_{OH}	$V_{CC} = 4.5$ V to 5.5 V,	$I_{OH} = -2$ mA	$V_{CC}-2$			$V_{CC}-2$			V
V_{OL}	$V_{CC} = 4.5$ V,	$I_{OL} = 20$ mA		0.35	0.5		0.35	0.5	V
I_I	$V_{CC} = 5.5$ V,	$V_I = 7$ V			0.1			0.1	mA
I_{IH}	$V_{CC} = 5.5$ V,	$V_I = 2.7$ V			20			20	μA
I_{IL}	$V_{CC} = 5.5$ V,	$V_I = 0.4$ V			−0.5			−0.5	mA
$I_O‡$	$V_{CC} = 5.5$ V,	$V_O = 2.25$ V	−30		−112	−30		−112	mA
I_{CCH}	$V_{CC} = 5.5$ V,	$V_I = 4.5$ V		5.8	9.3		5.8	9.3	mA
I_{CCL}	$V_{CC} = 5.5$ V,	$V_I = 0$ V		14.9	24		14.9	24	mA

† All typical values are at $V_{CC} = 5$ V, $T_A = 25$ °C.
‡ The output conditions hav been chosen to produce a current that closely approximates one half of the true short-circuit output current, I_{OS}.

switching characteristics (see Note 1)

PARAMETER	FROM (INPUT)	TO (OUTPUT)	$V_{CC} = 4.5$ V to 5.5 V, $C_L = 50$ pF, $R_L = 50$ Ω, $T_A = $ MIN to MAX				UNIT
			SN54AS08		SN74AS08		
			MIN	MAX	MIN	MAX	
t_{PLH}	A or B	Y	1	6.5	1	5.5	ns
t_{PHL}	A or B	Y	1	6.5	1	5.5	ns

NOTE 1: Load circuit and voltage waveforms are shown in Section 1.

TEXAS
INSTRUMENTS
POST OFFICE BOX 655012 • DALLAS, TEXAS 75265

ALS and AS Circuits

SN54AS32, SN74AS32
QUADRUPLE 2-INPUT POSITIVE-OR GATES

absolute maximum ratings over operating free-air temperature range (unless otherwise noted)

Supply voltage, V_{CC} . . . . 7 V
Input voltage . . . . 7 V
Operating free-air temperature range: SN54AS32 . . . . −55 °C to 125 °C
                                                                        SN74AS32 . . . . 0 °C to 70 °C
Storage temperature range . . . . −65 °C to 150 °C

recommended operating conditions

		SN54AS32			SN74AS32			UNIT
		MIN	NOM	MAX	MIN	NOM	MAX	
V_{CC}	Supply voltage	4.5	5	5.5	4.5	5	5.5	V
V_{IH}	High-level input voltage	2			2			V
V_{IL}	Low-level input voltage			0.8			0.8	mA
I_{OH}	High-level output current			−2			−2	mA
I_{OL}	Low--level output current			20			20	mA
T_A	Operating free-air temperature	−55		125	0		70	°C

electrical characteristics over recommended operating free-air temperature range (unless otherwise noted)

PARAMETER	TEST CONDITIONS		SN54AS32			SN74AS32			UNIT
			MIN	TYP†	MAX	MIN	TYP†	MAX	
V_{IK}	$V_{CC} = 4.5$ V,	$I_I = -18$ mA			−1.2			−1.2	V
V_{OH}	$V_{CC} = 4.5$ V to 5.5 V,	$I_{OH} = -2$ mA	$V_{CC}-2$			$V_{CC}-2$			V
V_{OL}	$V_{CC} = 4.5$ V,	$I_{OL} = 20$ mA		0.35	0.5		0.35	0.5	V
I_I	$V_{CC} = 5.5$ V,	$V_I = 7$ V			0.1			0.1	mA
I_{IH}	$V_{CC} = 5.5$ V,	$V_I = 2.7$ V			20			20	µA
I_{IL}	$V_{CC} = 5.5$ V,	$V_I = 0.4$ V			−0.5			−0.5	mA
I_O‡	$V_{CC} = 5.5$ V,	$V_O = 2.25$ V	−30		−112	−30		−112	mA
I_{CCH}	$V_{CC} = 5.5$ V,	$V_I = 4.5$ V		7.3	12		7.3	12	mA
I_{CCL}	$V_{CC} = 5.5$ V,	$V_I = 0$ V		16.5	26.6		16.5	26.6	mA

† All typical values are at $V_{CC} = 5$ V, $T_A = 25$ °C.
‡ The output conditions have been chosen to produce a current that closely approximates one half of the true short-circuit output current, I_{OS}.

switching characteristics (see Note 1)

PARAMETER	FROM (INPUT)	TO (OUTPUT)	$V_{CC} = 4.5$ V to 5.5 V, $C_L = 50$ pF, $R_L = 50$ Ω, $T_A =$ MIN to MAX				UNIT
			SN54AS32		SN74AS32		
			MIN	MAX	MIN	MAX	
t_{PLH}	A or B	Y	1	7.5	1	5.8	ns
t_{PHL}	A or B	Y	1	6.5	1	5.8	ns

NOTE 1: Load circuit and voltage waveforms are shown in Section 1.

ALS and AS Circuits

TEXAS INSTRUMENTS
POST OFFICE BOX 655012 • DALLAS, TEXAS 75265

SN54AS74, SN74AS74
DUAL D-TYPE POSITIVE-EDGE-TRIGGERED
FLIP-FLOPS WITH CLEAR AND PRESET

recommended operating conditions

			SN54AS74			SN74AS74			UNIT
			MIN	NOM	MAX	MIN	NOM	MAX	
V_{CC}	Supply voltage		4.5	5	5.5	4.5	5	5.5	V
V_{IH}	High-level input voltage		2			2			V
V_{IL}	Low-level input voltage				0.8			0.8	V
I_{OH}	High-level output current				-2			-2	mA
I_{OL}	Low-level output current				20			20	mA
f_{clock}	Clock frequency		0		90	0		105	MHz
t_w	Pulse duration	$\overline{PRE}$ or $\overline{CLR}$ low	4			4			ns
		CLK high	4			4			
		CLK low	5.5			5.5			
t_{su}	Setup time before CLK↑	Data	4.5			4.5			ns
		$\overline{PRE}$ or $\overline{CLR}$ inactive	2			2			
t_h	Hold time, data after CLK↑		0			0			ns
T_A	Operating free-air temperature		-55		125	0		70	°C

electrical characteristics over recommended operating free-air temperature range (unless otherwise noted)

PARAMETER		TEST CONDITIONS		SN54AS74			SN74AS74			UNIT
				MIN	TYP†	MAX	MIN	TYP†	MAX	
V_{IK}		$V_{CC} = 4.5$ V,	$I_I = -18$ mA			-1.2			-1.2	V
V_{OH}		$V_{CC} = 4.5$ V to 5.5 V,	$I_{OH} = -2$ mA	$V_{CC}-2$			$V_{CC}-2$			V
V_{OL}		$V_{CC} = 4.5$ V,	$I_{OL} = 20$ mA		0.25	0.5		0.25	0.5	V
I_I		$V_{CC} = 5.5$ V,	$V_I = 7$ V			0.1			0.1	mA
I_{IH}	CLK or D	$V_{CC} = 5.5$ V,	$V_I = 2.7$ V			20			20	μA
	$\overline{PRE}$ or $\overline{CLR}$					40			40	
I_{IL}	CLK or D	$V_{CC} = 5.5$ V,	$V_I = 0.4$ V			-0.5			-0.5	mA
	$\overline{PRE}$ or $\overline{CLR}$					-1.8			-1.8	
I_{IO}‡		$V_{CC} = 5.5$ V,	$V_O = 2.25$ V	-30		-112	-30		-112	mA
I_{CC}		$V_{CC} = 5.5$ V	See Note 1		10.5	16		10.5	16	mA

† All typical values are at $V_{CC} = 5$ V, $T_A = 25$°C.
‡ The output conditions have been chosen to produce a current that closely approximates one half of the true short-current output current, I_{OS}.
NOTE 1: I_{CC} is measured with D, CLK, and $\overline{PRE}$ grounded, then with D, CLK, and $\overline{CLR}$ grounded.

switching characteristics (see Note 2)

PARAMETER	FROM (INPUT)	TO (OUTPUT)	$V_{CC} = 4.5$ V to 5.5 V, $C_L = 50$ pF, $R_L = 500$ Ω, T_A = MIN to MAX				UNIT
			SN54AS74		SN74AS74		
			MIN	MAX	MIN	MAX	
f_{max}			90		105		MHz
t_{PLH}	$\overline{PRE}$ or $\overline{CLR}$	Q or $\overline{Q}$	3	8.5	3	7.5	ns
t_{PHL}			3.5	11.5	3.5	10.5	
t_{PLH}	CLK	Q or $\overline{Q}$	3.5	9	3.5	8	ns
t_{PHL}			4.5	10.5	4.5	9	

NOTE 2: Load circuit and voltage waveforms are shown in Section 1.

TEXAS INSTRUMENTS
POST OFFICE BOX 655012 • DALLAS, TEXAS 75265

ALS and AS Circuits

Reprinted by permission of Texas Instruments.

recommended operating conditions

			SN54AS109			SN74AS109			UNIT
			MIN	NOM	MAX	MIN	NOM	MAX	
V_{CC}	Supply voltage		4.5	5	5.5	4.5	5	5.5	V
V_{IH}	High-level input voltage		2			2			V
V_{IL}	Low-level input voltage				0.8			0.8	V
I_{OH}	High-level output current				−2			−2	mA
I_{OL}	Low-level output current				20			20	mA
f_{clock}	Clock frequency		0		90	0		105	MHz
t_w	Pulse duration	$\overline{PRE}$ or $\overline{CLR}$ low	4			4			ns
		CLK high	4			4			
		CLK low	5.5			5.5			
t_{su}	Setup time before CLK↑	Data	5.5			5.5			ns
		$\overline{PRE}$ or $\overline{CLR}$ inactive	2			2			
t_h	Hold time, data after CLK↑		0			0			ns
T_A	Operating free-air temperature		−55		125	0		70	°C

electrical characteristics over recommended operating free-air temperature range (unless otherwise noted)

PARAMETER		TEST CONDITIONS		SN54AS109			SN74AS109			UNIT
				MIN	TYP†	MAX	MIN	TYP†	MAX	
V_{IK}		V_{CC} = 4.5 V,	I_I = −18 mA			−1.2			−1.2	V
V_{OH}		V_{CC} = 4.5 V to 5.5 V,	I_{OH} = −2 mA	V_{CC}−2			V_{CC}−2			V
V_{OL}		V_{CC} = 4.5 V,	I_{OL} = 20 mA		0.25	0.5		0.25	0.5	V
I_I		V_{CC} = 5.5 V,	V_I = 7 V			0.1			0.1	mA
I_{IH}	CLK, J or $\overline{K}$	V_{CC} = 5.5 V,	V_I = 2.7 V			20			20	μA
	$\overline{PRE}$ or $\overline{CLR}$					40			40	
I_{IL}	CLK, J or $\overline{K}$	V_{CC} = 5.5 V,	V_I = 0.4 V			−0.5			0.5	mA
	$\overline{PRE}$ or $\overline{CLR}$					−1.8			−1.8	
I_O‡		V_{CC} = 5.5 V,	V_O = 2.25 V	−30		−112	−30		−112	mA
I_{CC}		V_{CC} = 5.5 V,	See Note 1		11.5	17		11.5	17	mA

†All typical values are at V_{CC} = 5 V, T_A = 25 °C.
‡The output conditions have been chosen to produce a current that closely approximates one half of the true short-circuit output current, I_{OS}.
NOTE 1: I_{CC} is measured with J, $\overline{K}$, CLK, and $\overline{PRE}$ grounded, then with J, $\overline{K}$, CLK, and $\overline{CLR}$ grounded.

switching characteristics (see Note 2)

PARAMETER	FROM (INPUT)	TO (OUTPUT)	V_{CC} = 4.5 V to 5.5 V, C_L = 50 pF, R_L = 500 Ω, T_A = MIN to MAX				UNIT
			SN54AS109		SN74AS109		
			MIN	MAX	MIN	MAX	
f_{max}			90		105		MHz
t_{PLH}	$\overline{PRE}$ or $\overline{CLR}$	Q or $\overline{Q}$	3	9	3	8	ns
t_{PHL}			3.5	11.5	3.5	10.5	
t_{PLH}	CLK	Q or $\overline{Q}$	3.5	10	3.5	9	ns
t_{PHL}			4.5	10.5	4.5	9	

NOTE 2: Load circuit and voltage waveforms are shown in Section 1.

ALS and AS Circuits

TEXAS
INSTRUMENTS
POST OFFICE BOX 655012 • DALLAS, TEXAS 75265

SN54HC00, SN74HC00
QUADRUPLE 2-INPUT POSITIVE-NAND GATES

HCMOS Devices

absolute maximum ratings over operating free-air temperature range[†]

Supply voltage, V_{CC} .... -0.5 V to 7 V
Input clamp current, I_{IK} ($V_I < 0$ or $V_I > V_{CC}$) .... ± 20 mA
Output clamp current, I_{OK} ($V_O < 0$ or $V_O > V_{CC}$) .... ± 20 mA
Continuous output current, I_O ($V_O = 0$ to V_{CC}) .... ± 25 mA
Continuous current through V_{CC} or GND pins .... ± 50 mA
Lead temperature 1,6 mm (1/16 in) from case for 60 s: FK or J package .... 300°C
Lead temperature 1,6 mm (1/16 in) from case for 10 s: D or N package .... 260°C
Storage temperature range .... -65°C to 150°C

[†] Stresses beyond those listed under "absolute maximum ratings" may cause permanent damage to the device. These are stress ratings only, and functional operation of the device at these or any other conditions beyond those indicated under "recommended operating conditions" is not implied. Exposure to absolute-maximum-rated conditions for extended periods may affect device reliability.

recommended operating conditions

			SN54HC00			SN74HC00			UNIT
			MIN	NOM	MAX	MIN	NOM	MAX	
V_{CC}	Supply voltage		2	5	6	2	5	6	V
V_{IH}	High-level input voltage	V_{CC} = 2 V	1.5			1.5			V
		V_{CC} = 4.5 V	3.15			3.15			
		V_{CC} = 6 V	4.2			4.2			
V_{IL}	Low-level input voltage	V_{CC} = 2 V	0		0.3	0		0.3	V
		V_{CC} = 4.5 V	0		0.9	0		0.9	
		V_{CC} = 6 V	0		1.2	0		1.2	
V_I	Input voltage		0		V_{CC}	0		V_{CC}	V
V_O	Output voltage		0		V_{CC}	0		V_{CC}	V
t_t	Input transition (rise and fall) times	V_{CC} = 2 V	0		1000	0		1000	ns
		V_{CC} = 4.5 V	0		500	0		500	
		V_{CC} = 6 V	0		400	0		400	
T_A	Operating free-air temperature		-55		125	-40		85	°C

electrical characteristics over recommended operating free-air temperature range (unless otherwise noted)

PARAMETER	TEST CONDITIONS	V_{CC}	T_A = 25°C			SN54HC00		SN74HC00		UNIT
			MIN	TYP	MAX	MIN	MAX	MIN	MAX	
V_{OH}	$V_I = V_{IH}$ or V_{IL}, $I_{OH} = -20$ μA	2 V	1.9	1.998		1.9		1.9		V
		4.5 V	4.4	4.499		4.4		4.4		
		6 V	5.9	5.999		5.9		5.9		
	$V_I = V_{IH}$ or V_{IL}, $I_{OH} = -4$ mA	4.5 V	3.98	4.30		3.7		3.84		
	$V_I = V_{IH}$ or V_{IL}, $I_{OH} = -5.2$ mA	6 V	5.48	5.80		5.2		5.34		
V_{OL}	$V_I = V_{IH}$ or V_{IL}, $I_{OL} = 20$ μA	2 V		0.002	0.1		0.1		0.1	V
		4.5 V		0.001	0.1		0.1		0.1	
		6 V		0.001	0.1		0.1		0.1	
	$V_I = V_{IH}$ or V_{IL}, $I_{OL} = 4$ mA	4.5 V		0.17	0.26		0.4		0.33	
	$V_I = V_{IH}$ or V_{IL}, $I_{OL} = 5.2$ mA	6 V		0.15	0.26		0.4		0.33	
I_I	$V_I = V_{CC}$ or 0	6 V		± 0.1	± 100		± 1000		± 1000	nA
I_{CC}	$V_I = V_{CC}$ or 0, $I_O = 0$	6 V			2		40		20	μA
C_i		2 to 6 V		3	10		10		10	pF

switching characteristics over recommended operating free-air temperature range (unless otherwise noted), C_L = 50 pF (see Note 1)

PARAMETER	FROM (INPUT)	TO (OUTPUT)	V_{CC}	T_A = 25°C			SN54HC00		SN74HC00		UNIT
				MIN	TYP	MAX	MIN	MAX	MIN	MAX	
t_{pd}	A or B	Y	2 V		45	90		135		115	ns
			4.5 V		9	18		27		23	
			6 V		8	15		23		20	
t_t		Y	2 V		38	75		110		95	ns
			4.5 V		8	15		22		19	
			6 V		6	13		19		16	

C_{pd}	Power dissipation capacitance per gate	No load, T_A = 25°C	20 pF typ

NOTE 1: Load circuit and voltage waveforms are shown in Section 1.

TEXAS INSTRUMENTS
POST OFFICE BOX 655012 • DALLAS, TEXAS 75265

Reprinted by permission of Texas Instruments.

SN54HC02, SN74HC02
QUADRUPLE 2-INPUT POSITIVE-NOR GATES

absolute maximum ratings over operating free-air temperature range[†]

Supply voltage, V_{CC}	-0.5 V to 7 V
Input clamp current, I_{IK} ($V_I < 0$ or $V_I > V_{CC}$)	± 20 mA
Output clamp current, I_{OK} ($V_O < 0$ or $V_O > V_{CC}$)	± 20 mA
Continuous output current, I_O ($V_O = 0$ to V_{CC})	± 25 mA
Continuous current through V_{CC} or GND pins	± 50 mA
Lead temperature 1,6 mm (1/16 in) from case for 60 s: FK or J package	$300\,°C$
Lead temperature 1,6 mm (1/16 in) from case for 10 s: D or N package	$260\,°C$
Storage temperature range	$-65\,°C$ to $150\,°C$

[†] Stresses beyond those listed under "absolute maximum ratings" may cause permanent damage to the device. These are stress ratings only, and functional operation of the device at these or any other conditions beyond those indicated under "recommended operating conditions" is not implied. Exposure to absolute-maximum-rated conditions for extended periods may affect device reliability.

recommended operating conditions

			SN54HC02			SN74HC02			UNIT
			MIN	NOM	MAX	MIN	NOM	MAX	
V_{CC}	Supply voltage		2	5	6	2	5	6	V
V_{IH}	High-level input voltage	$V_{CC} = 2$ V	1.5			1.5			V
		$V_{CC} = 4.5$ V	3.15			3.15			
		$V_{CC} = 6$ V	4.2			4.2			
V_{IL}	Low-level input voltage	$V_{CC} = 2$ V	0		0.3	0		0.3	V
		$V_{CC} = 4.5$ V	0		0.9	0		0.9	
		$V_{CC} = 6$ V	0		1.2	0		1.2	
V_I	Input voltage		0		V_{CC}	0		V_{CC}	V
V_O	Output voltage		0		V_{CC}	0		V_{CC}	V
t_t	Input transition (rise and fall) times	$V_{CC} = 2$ V	0		1000	0		1000	ns
		$V_{CC} = 4.5$ V	0		500	0		500	
		$V_{CC} = 6$ V	0		400	0		400	
T_A	Operating free-air temperature		-55		125	-40		85	$°C$

electrical characteristics over recommended operating free-air temperature range (unless otherwise noted)

PARAMETER	TEST CONDITIONS		V_{CC}	$T_A = 25\,°C$			SN54HC02		SN74HC02		UNIT
				MIN	TYP	MAX	MIN	MAX	MIN	MAX	
V_{OH}	$V_I = V_{IH}$ or V_{IL}, $I_{OH} = -20\ \mu A$		2 V	1.9	1.998		1.9		1.9		V
			4.5 V	4.4	4.499		4.4		4.4		
			6 V	5.9	5.999		5.9		5.9		
	$V_I = V_{IH}$ or V_{IL}, $I_{OH} = -4$ mA		4.5 V	3.98	4.30		3.7		3.84		
	$V_I = V_{IH}$ or V_{IL}, $I_{OH} = -5.2$ mA		6 V	5.48	5.80		5.2		5.34		
V_{OL}	$V_I = V_{IH}$ or V_{IL}, $I_{OL} = 20\ \mu A$		2 V		0.002	0.1		0.1		0.1	V
			4.5 V		0.001	0.1		0.1		0.1	
			6 V		0.001	0.1		0.1		0.1	
	$V_I = V_{IH}$ or V_{IL}, $I_{OL} = 4$ mA		4.5 V		0.17	0.26		0.4		0.33	
	$V_I = V_{IH}$ or V_{IL}, $I_{OL} = 5.2$ mA		6 V		0.15	0.26		0.4		0.33	
I_I	$V_I = 0$ or V_{CC}		6 V		± 0.1	± 100		± 1000		± 1000	nA
I_{CC}	$V_I = V_{CC}$ or 0, $I_O = 0$		6 V			2		40		20	μA
C_i			2 to 6 V		3	10		10		10	pF

switching characteristics over recommended operating free-air temperature range (unless otherwise noted), $C_L = 50$ pF (see Note 1)

PARAMETER	FROM (INPUT)	TO (OUTPUT)	V_{CC}	$T_A = 25\,°C$			SN54HC02		SN74HC02		UNIT
				MIN	TYP	MAX	MIN	MAX	MIN	MAX	
t_{pd}	A or B	Y	2 V		45	90		135		115	ns
			4.5 V		9	18		27		23	
			6 V		8	15		23		20	
t_t		Y	2 V		38	75		110		95	ns
			4.5 V		8	15		22		19	
			6 V		6	13		19		16	

C_{pd}	Power dissipation capacitance per gate	No load, $T_A = 25\,°C$	22 pF typ

NOTE 1: Load circuit and voltage waveforms are shown in Section 1.

TEXAS INSTRUMENTS

POST OFFICE BOX 655012 • DALLAS, TEXAS 75265

Reprinted by permission of Texas Instruments.

SN54HC04, SN74HC04
HEX INVERTERS

absolute maximum ratings over operating free-air temperature range[†]

Supply voltage, V_{CC} .... -0.5 V to 7 V
Input clamp current, I_{IK} ($V_I < 0$ or $V_I > V_{CC}$) .... ± 20 mA
Output clamp current, I_{OK} ($V_O < 0$ or $V_O > V_{CC}$) .... ± 20 mA
Continuous output current, I_O ($V_O = 0$ to V_{CC}) .... ± 25 mA
Continuous current through V_{CC} or GND pins .... ± 50 mA
Lead temperature 1,6 mm (1/16 in) from case for 60 s: FK or J package .... 300 °C
Lead temperature 1,6 mm (1/16 in) from case for 10 s: D or N package .... 260 °C
Storage temperature range .... -65 °C to 150 °C

[†] Stresses beyond those listed under "absolute maximum ratings" may cause permanent damage to the device. These are stress ratings only, and functional operation of the device at these or any other conditions beyond those indicated under "recommended operating conditions" is not implied. Exposure to absolute-maximum-rated conditions for extended periods may affect device reliability.

recommended operating conditions

			SN54HC04			SN74HC04			UNIT
			MIN	NOM	MAX	MIN	NOM	MAX	
V_{CC}	Supply voltage		2	5	6	2	5	6	V
V_{IH}	High-level input voltage	$V_{CC} = 2$ V	1.5			1.5			V
		$V_{CC} = 4.5$ V	3.15			3.15			
		$V_{CC} = 6$ V	4.2			4.2			
V_{IL}	Low-level input voltage	$V_{CC} = 2$ V	0		0.3	0		0.3	V
		$V_{CC} = 4.5$ V	0		0.9	0		0.9	
		$V_{CC} = 6$ V	0		1.2	0		1.2	
V_I	Input voltage		0		V_{CC}	0		V_{CC}	V
V_O	Output voltage		0		V_{CC}	0		V_{CC}	V
t_t	Input transition (rise and fall) times	$V_{CC} = 2$ V	0		1000	0		1000	ns
		$V_{CC} = 4.5$ V	0		500	0		500	
		$V_{CC} = 6$ V	0		400	0		400	
T_A	Operating free-air temperature		-55		125	-40		85	°C

electrical characteristics over recommended operating free-air temperature range (unless otherwise noted)

PARAMETER	TEST CONDITIONS	V_{CC}	$T_A = 25$ °C			SN54HC04		SN74HC04		UNIT
			MIN	TYP	MAX	MIN	MAX	MIN	MAX	
V_{OH}	$V_I = V_{IH}$ or V_{IL}, $I_{OH} = -20$ μA	2 V	1.9	1.998		1.9		1.9		V
		4.5 V	4.4	4.499		4.4		4.4		
		6 V	5.9	5.999		5.9		5.9		
	$V_I = V_{IH}$ or V_{IL}, $I_{OH} = -4$ mA	4.5 V	3.98	4.30		3.7		3.84		
	$V_I = V_{IH}$ or V_{IL}, $I_{OH} = -5.2$ mA	6 V	5.48	5.80		5.2		5.34		
V_{OL}	$V_I = V_{IH}$ or V_{IL}, $I_{OL} = 20$ μA	2 V		0.002	0.1		0.1		0.1	V
		4.5 V		0.001	0.1		0.1		0.1	
		6 V		0.001	0.1		0.1		0.1	
	$V_I = V_{IH}$ or V_{IL}, $I_{OL} = 4$ mA	4.5 V		0.17	0.26		0.4		0.33	
	$V_I = V_{IH}$ or V_{IL}, $I_{OL} = 5.2$ mA	6 V		0.15	0.26		0.4		0.33	
I_I	$V_I = 0$ or V_{CC}	6 V		± 0.1	± 100		± 1000		± 1000	nA
I_{CC}	$V_I = V_{CC}$ or 0, $I_O = 0$	6 V			2		40		20	μA
C_i		2 to 6 V		3	10		10		10	pF

switching characteristics over recommended operating free-air temperature range (unless otherwise noted), $C_L = 50$ pF (see Note 1)

PARAMETER	FROM (INPUT)	TO (OUTPUT)	V_{CC}	$T_A = 25$ °C			SN54HC04		SN74HC04		UNIT
				MIN	TYP	MAX	MIN	MAX	MIN	MAX	
t_{pd}	A	Y	2 V		45	95		145		120	ns
			4.5 V		9	19		29		24	
			6 V		8	16		25		20	
t_t		Y	2 V		38	75		110		95	ns
			4.5 V		8	15		22		19	
			6 V		6	13		19		16	

C_{pd}	Power dissipation capacitance per inverter	No load, $T_A = 25$ °C	20 pF typ

NOTE 1: Load circuit and voltage waveforms are shown in Section 1.

TEXAS
INSTRUMENTS
POST OFFICE BOX 655012 • DALLAS, TEXAS 75265

Reprinted by permission of Texas Instruments.

SN54HC05, SN74HC05
HEX INVERTERS WITH OPEN-DRAIN OUTPUTS

absolute maximum ratings over operating free-air temperature range†

Supply voltage, V_{CC} . . . . −0.5 V to 7 V
Input clamp current, I_{IK} ($V_I < 0$ or $V_I > V_{CC}$) . . . . ±20 mA
Output clamp current, I_{OK} ($V_O < 0$ or $V_O > V_{CC}$) . . . . ±20 mA
Continuous output current, I_O ($V_O = 0$ to V_{CC}) . . . . ±25 mA
Continuous current through V_{CC} or GND pins . . . . ±50 mA
Lead temperature 1,6 mm (1/16 in) from case for 60 s: FK or J package . . . . 300°C
Lead temperature 1,6 mm (1/16 in) from case for 10 s: D or N package . . . . 260°C
Storage temperature range . . . . −65°C to 150°C

† Stresses beyond those listed under "absolute maximum ratings" may cause permanent damage to the device. These are stress ratings only, and functional operation of the device at these or any other conditions beyond those indicated under "recommended operating conditions" is not implied. Exposure to absolute-maximum-rated conditions for extended periods may affect device reliability.

recommended operating conditions

			SN54HC05			SN74HC05			UNIT
			MIN	NOM	MAX	MIN	NOM	MAX	
V_{CC}	Supply voltage		2	5	6	2	5	6	V
V_{IH}	High-level input voltage	$V_{CC} = 2$ V	1.5			1.5			V
		$V_{CC} = 4.5$ V	3.15			3.15			
		$V_{CC} = 6$ V	4.2			4.2			
V_{IL}	Low-level input voltage	$V_{CC} = 2$ V	0		0.3	0		0.3	V
		$V_{CC} = 4.5$ V	0		0.9	0		0.9	
		$V_{CC} = 6$ V	0		1.2	0		1.2	
V_I	Input voltage		0		V_{CC}	0		V_{CC}	V
V_O	Output voltage		0		V_{CC}	0		V_{CC}	V
t_t	Input transition (rise and fall) times	$V_{CC} = 2$ V	0		1000	0		1000	ns
		$V_{CC} = 4.5$ V	0		500	0		500	
		$V_{CC} = 6$ V	0		400	0		400	
T_A	Operating free-air temperature		−55		125	−40		85	°C

electrical characteristics over recommended operating free-air temperature range (unless otherwise noted)

PARAMETER	TEST CONDITIONS	V_{CC}	$T_A = 25$°C			SN54HC05		SN74HC05		UNIT
			MIN	TYP	MAX	MIN	MAX	MIN	MAX	
I_{OH}	$V_I = V_{IH}$ or V_{IL}, $V_O = V_{CC}$	6 V		0.01	0.5		10		5	μA
V_{OL}	$V_I = V_{IH}$ or V_{IL}, $I_{OL} = 20$ μA	2 V		0.002	0.1		0.1		0.1	V
		4.5 V		0.001	0.1		0.1		0.1	
		6 V		0.001	0.1		0.1		0.1	
	$V_I = V_{IH}$ or V_{IL}, $I_{OL} = 4$ mA	4.5 V		0.17	0.26		0.4		0.33	
	$V_I = V_{IH}$ or V_{IL}, $I_{OL} = 5.2$ mA	6 V		0.15	0.26		0.4		0.33	
I_I	$V_I = V_{CC}$ or 0	6 V		±0.1	±100		±1000		±1000	nA
I_{CC}	$V_I = V_{CC}$ or 0, $I_O = 0$	6 V			2		40		20	μA
C_i		2 to 6 V		3	10		10		10	pF

switching characteristics over recommended operating free-air temperature range (unless otherwise noted), $C_L = 50$ pF (see Note 1)

PARAMETER	FROM (INPUT)	TO (OUTPUT)	V_{CC}	$T_A = 25$°C			SN54HC05		SN74HC05		UNIT
				MIN	TYP	MAX	MIN	MAX	MIN	MAX	
t_{PLH}	A	Y	2 V		60	115		175		145	ns
			4.5 V		13	23		35		29	
			6 V		10	20		30		25	
t_{PHL}	A	Y	2 V		45	85		130		105	ns
			4.5 V		9	17		26		21	
			6 V		8	14		22		18	
t_f		Y	2 V		38	75		110		95	ns
			4.5 V		8	15		22		19	
			6 V		6	13		19		16	

C_{pd}	Power dissipation capacitance per inverter	No load, $T_A = 25$°C	20 pF typ

NOTE 1: Load circuit and voltage waveforms are shown in Section 1.

TEXAS
INSTRUMENTS
POST OFFICE BOX 655012 • DALLAS, TEXAS 75265

Reprinted by permission of Texas Instruments.

HCMOS Devices

SN54HC08, SN74HC08
QUADRUPLE 2-INPUT POSITIVE-AND GATES

HCMOS Devices

absolute maximum ratings over operating free-air temperature range[†]

Supply voltage, V_{CC} . . . . -0.5 V to 7 V
Input clamp current, I_{IK} ($V_I < 0$ or $V_I > V_{CC}$) . . . . ± 20 mA
Output clamp current, I_{OK} ($V_O < 0$ or $V_O > V_{CC}$) . . . . ± 20 mA
Continuous output current, I_O ($V_O = 0$ to V_{CC}) . . . . ± 25 mA
Continuous current through V_{CC} or GND pins . . . . ± 50 mA
Lead temperature 1,6 mm (1/16 in) from case for 60 s: FK or J package . . . . 300°C
Lead temperature 1,6 mm (1/16 in) from case for 10 s: D or N package . . . . 260°C
Storage temperature range . . . . -65°C to 150°C

[†] Stresses beyond those listed under "absolute maximum ratings" may cause permanent damage to the device. These are stress ratings only, and functional operation of the device at these or any other conditions beyond those indicated under "recommended operating conditions" is not implied. Exposure to absolute-maximum-rated conditions for extended periods may affect device reliability.

recommended operating conditions

			SN54HC08			SN74HC08			UNIT
			MIN	NOM	MAX	MIN	NOM	MAX	
V_{CC}	Supply voltage		2	5	6	2	5	6	V
V_{IH}	High-level input voltage	$V_{CC} = 2$ V	1.5			1.5			V
		$V_{CC} = 4.5$ V	3.15			3.15			
		$V_{CC} = 6$ V	4.2			4.2			
V_{IL}	Low-level input voltage	$V_{CC} = 2$ V	0		0.3	0		0.3	V
		$V_{CC} = 4.5$ V	0		0.9	0		0.9	
		$V_{CC} = 6$ V	0		1.2	0		1.2	
V_I	Input voltage		0		V_{CC}	0		V_{CC}	V
V_O	Output voltage		0		V_{CC}	0		V_{CC}	V
t_t	Input transition (rise and fall) times	$V_{CC} = 2$ V	0		1000	0		1000	ns
		$V_{CC} = 4.5$ V	0		500	0		500	
		$V_{CC} = 6$ V	0		400	0		400	
T_A	Operating free-air temperature		-55		125	-40		85	°C

electrical characteristics over recommended operating free-air temperature range (unless otherwise noted)

PARAMETER	TEST CONDITIONS	V_{CC}	$T_A = 25$°C			SN54HC08		SN74HC08		UNIT
			MIN	TYP	MAX	MIN	MAX	MIN	MAX	
V_{OH}	$V_I = V_{IH}$ or V_{IL}, $I_{OH} = -20$ μA	2 V	1.9	1.998		1.9		1.9		V
		4.5 V	4.4	4.499		4.4		4.4		
		6 V	5.9	5.999		5.9		5.9		
	$V_I = V_{IH}$ or V_{IL}, $I_{OH} = -4$ mA	4.5 V	3.98	4.30		3.7		3.84		
	$V_I = V_{IH}$ or V_{IL}, $I_{OH} = -5.2$ mA	6 V	5.48	5.80		5.2		5.34		
V_{OL}	$V_I = V_{IH}$ or V_{IL}, $I_{OL} = 20$ μA	2 V		0.002	0.1		0.1		0.1	V
		4.5 V		0.001	0.1		0.1		0.1	
		6 V		0.001	0.1		0.1		0.1	
	$V_I = V_{IH}$ or V_{IL}, $I_{OL} = 4$ mA	4.5 V		0.17	0.26		0.4		0.33	
	$V_I = V_{IH}$ or V_{IL}, $I_{OL} = 5.2$ mA	6 V		0.15	0.26		0.4		0.33	
I_I	$V_I = V_{CC}$ or 0	6 V		± 0.1	± 100		± 1000		± 1000	nA
I_{CC}	$V_I = V_{CC}$ or 0, $I_O = 0$	6 V			2		40		20	μA
C_i		2 to 6 V		3	10		10		10	pF

switching characteristics over recommended operating free-air temperature range (unless otherwise noted), $C_L = 50$ pF (see Note 1)

PARAMETER	FROM (INPUT)	TO (OUTPUT)	V_{CC}	$T_A = 25$°C			SN54HC08		SN74HC08		UNIT
				MIN	TYP	MAX	MIN	MAX	MIN	MAX	
t_{pd}	A or B	Y	2 V		50	100		150		125	ns
			4.5 V		10	20		30		25	
			6 V		8	17		25		21	
t_t		Y	2 V		38	75		110		95	ns
			4.5 V		8	15		22		19	
			6 V		6	13		19		16	

C_{pd}	Power dissipation capacitance per gate	No load, $T_A = 25$°C	20 pF typ

NOTE 1: Load circuit and voltage waveforms are shown in Section 1.

TEXAS INSTRUMENTS
POST OFFICE BOX 655012 • DALLAS, TEXAS 75265

Reprinted by permission of Texas Instruments.

SN54HC32, SN74HC32
QUADRUPLE 2-INPUT POSITIVE-OR GATES

HCMOS Devices

absolute maximum ratings over operating free-air temperature range[†]

Supply voltage, V_{CC}	-0.5 V to 7 V
Input clamp current, I_{IK} ($V_I < 0$ or $V_I > V_{CC}$)	± 20 mA
Output clamp current, I_{OK} ($V_O < 0$ or $V_O > V_{CC}$)	± 20 mA
Continuous output current, I_O ($V_O = 0$ to V_{CC})	± 25 mA
Continuous current through V_{CC} or GND pins	± 50 mA
Lead temperature 1,6 mm (1/16 in) from case for 60 s: FK or J package	300 °C
Lead temperature 1,6 mm (1/16 in) from case for 10 s: D or N package	260 °C
Storage temperature range	-65 °C to 150 °C

[†] Stresses beyond those listed under "absolute maximum ratings" may cause permanent damage to the device. These are stress ratings only, and functional operation of the device at these or any other conditions beyond those indicated under "recommended operating conditions" is not implied. Exposure to absolute-maximum-rated conditions for extended periods may affect device reliability.

recommended operating conditions

			SN54HC32			SN74HC32			UNIT
			MIN	NOM	MAX	MIN	NOM	MAX	
V_{CC}	Supply voltage		2	5	6	2	5	6	V
V_{IH}	High-level input voltage	$V_{CC} = 2$ V	1.5			1.5			V
		$V_{CC} = 4.5$ V	3.15			3.15			
		$V_{CC} = 6$ V	4.2			4.2			
V_{IL}	Low-level input voltage	$V_{CC} = 2$ V	0		0.3	0		0.3	V
		$V_{CC} = 4.5$ V	0		0.9	0		0.9	
		$V_{CC} = 6$ V	0		1.2	0		1.2	
V_I	Input voltage		0		V_{CC}	0		V_{CC}	V
V_O	Output voltage		0		V_{CC}	0		V_{CC}	V
t_t	Input transition (rise and fall) times	$V_{CC} = 2$ V	0		1000	0		1000	ns
		$V_{CC} = 4.5$ V	0		500	0		500	
		$V_{CC} = 6$ V	0		400	0		400	
T_A	Operating free-air temperature		-55		125	-40		85	°C

electrical characteristics over recommended operating free-air temperature range (unless otherwise noted)

PARAMETER	TEST CONDITIONS	V_{CC}	$T_A = 25°C$			SN54HC32		SN74HC32		UNIT
			MIN	TYP	MAX	MIN	MAX	MIN	MAX	
V_{OH}	$V_I = V_{IH}$ or V_{IL}, $I_{OH} = -20$ μA	2 V	1.9	1.998		1.9		1.9		V
		4.5 V	4.4	4.499		4.4		4.4		
		6 V	5.9	5.999		5.9		5.9		
	$V_I = V_{IH}$ or V_{IL}, $I_{OH} = -4$ mA	4.5 V	3.98	4.30		3.7		3.84		
	$V_I = V_{IH}$ or V_{IL}, $I_{OH} = -5.2$ mA	6 V	5.48	5.80		5.2		5.34		
V_{OL}	$V_I = V_{IH}$ or V_{IL}, $I_{OL} = 20$ μA	2 V		0.002	0.1		0.1		0.1	V
		4.5 V		0.001	0.1		0.1		0.1	
		6 V		0.001	0.1		0.1		0.1	
	$V_I = V_{IH}$ or V_{IL}, $I_{OL} = 4$ mA	4.5 V		0.17	0.26		0.4		0.33	
	$V_I = V_{IH}$ or V_{IL}, $I_{OL} = 5.2$ mA	6 V		0.15	0.26		0.4		0.33	
I_I	$V_I = V_{CC}$ or 0	6 V		± 0.1	± 100		± 1000		± 1000	nA
I_{CC}	$V_I = V_{CC}$ or 0, $I_O = 0$	6 V			2		40		20	μA
C_i		2 to 6 V		3	10		10		10	pF

switching characteristics over recommended operating free-air temperature range (unless otherwise noted), $C_L = 50$ pF (see Note 1)

PARAMETER	FROM (INPUT)	TO (OUTPUT)	V_{CC}	$T_A = 25°C$			SN54HC32		SN74HC32		UNIT
				MIN	TYP	MAX	MIN	MAX	MIN	MAX	
t_{pd}	A or B	Y	2 V		50	100		150		125	ns
			4.5 V		10	20		30		25	
			6 V		8	17		25		21	
t_t		Y	2 V		38	75		110		95	ns
			4.5 V		8	15		22		19	
			6 V		6	13		19		16	

C_{pd}	Power dissipation capacitance per gate	No load, $T_A = 25°C$	20 pF typ

NOTE 1: Load circuit and voltage waveforms are shown in Section 1.

POST OFFICE BOX 655012 • DALLAS, TEXAS 75265

Reprinted by permission of Texas Instruments.

SN54HC74, SN74HC74
DUAL D-TYPE POSITIVE-EDGE-TRIGGERED
FLIP-FLOPS WITH CLEAR AND PRESET

HCMOS Devices

absolute maximum ratings over operating free-air temperature range[†]

Supply voltage, V_{CC} . . . . −0.5 V to 7 V
Input clamp current, I_{IK} ($V_I < 0$ or $V_I > V_{CC}$) . . . . ±20 mA
Output clamp current, I_{OK} ($V_O < 0$ or $V_O > V_{CC}$ . . . . ±20 mA
Continuous output current, I_O ($V_O = 0$ to V_{CC}) . . . . ±25 mA
Continuous current through V_{CC} or GND pins . . . . ±50 mA
Lead temperature 1,6 mm (1/16 in) from case for 60 s: FK or J package . . . . 300°C
Lead temperature 1,6 mm (1/16 in) from case for 10 s: D or N package . . . . 260°C
Storage temperature range . . . . −65°C to 150°C

[†] Stresses beyond those listed under ''absolute maximum ratings'' may cause permanent damage to the device. These are stress ratings only, and functional operation of the device at these or any other conditions beyond those indicated under ''recommended operating conditions'' is not implied. Exposure to absolute-maximum-rated conditions for extended periods may affect device reliability.

recommended operating conditions

			SN54HC74			SN74HC74			UNIT
			MIN	NOM	MAX	MIN	NOM	MAX	
V_{CC}	Supply voltage		2	5	6	2	5	6	V
V_{IH}	High-level input voltage	$V_{CC} = 2$ V	1.5			1.5			V
		$V_{CC} = 4.5$ V	3.15			3.15			
		$V_{CC} = 6$ V	4.2			4.2			
V_{IL}	Low-level input voltage	$V_{CC} = 2$ V	0		0.3	0		0.3	V
		$V_{CC} = 4.5$ V	0		0.9	0		0.9	
		$V_{CC} = 6$ V	0		1.2	0		1.2	
V_I	Input voltage		0		V_{CC}	0		V_{CC}	V
V_O	Output voltage		0		V_{CC}	0		V_{CC}	V
t_t	Input transition (rise and fall) times	$V_{CC} = 2$ V	0		1000	0		1000	ns
		$V_{CC} = 4.5$ V	0		500	0		500	
		$V_{CC} = 6$ V	0		400	0		400	
T_A	Operating free-air temperature		−55		125	−40		85	°C

electrical characteristics over recommended operating free-air temperature range (unless otherwise noted)

PARAMETER	TEST CONDITIONS	V_{CC}	$T_A = 25°C$			SN54HC74		SN74HC74		UNIT
			MIN	TYP	MAX	MIN	MAX	MIN	MAX	
V_{OH}	$V_I = V_{IH}$ or V_{IL}, $I_{OH} = -20$ μA	2 V	1.9	1.998		1.9		1.9		V
		4.5 V	4.4	4.499		4.4		4.4		
		6 V	5.9	5.999		5.9		5.9		
	$V_I = V_{IH}$ or V_{IL}, $I_{OH} = -4$ mA	4.5 V	3.98	4.30		3.7		3.84		
	$V_I = V_{IH}$ or V_{IL}, $I_{OH} = -5.2$ mA	6 V	5.48	5.80		5.2		5.34		
V_{OL}	$V_I = V_{IH}$ or V_{IL}, $I_{OL} = 20$ μA	2 V		0.002	0.1		0.1		0.1	V
		4.5 V		0.001	0.1		0.1		0.1	
		6 V		0.001	0.1		0.1		0.1	
	$V_I = V_{IH}$ or V_{IL}, $I_{OL} = 4$ mA	4.5 V		0.17	0.26		0.4		0.33	
	$V_I = V_{IH}$ or V_{IL}, $I_{OL} = 5.2$ mA	6 V		0.15	0.26		0.4		0.33	
I_I	$V_I = 0$ or V_{CC}	6 V		±0.1	±100		±1000		±1000	nA
I_{CC}	$V_I = 0$ or V_{CC}, $I_O = 0$	6 V			4		80		40	μA
C_i		2 to 6 V		3	10		10		10	pF

TEXAS
INSTRUMENTS

POST OFFICE BOX 655012 • DALLAS, TEXAS 75265

HCMOS Devices

timing requirements over recommended operating free-air temperature range (unless otherwise noted)

		V_{CC}	$T_A = 25°C$		SN54HC74		SN74HC74		UNIT
			MIN	MAX	MIN	MAX	MIN	MAX	
f_{clock} Clock frequency		2 V	0	6	0	4.2	0	5	
		4.5 V	0	31	0	21	0	25	MHz
		6 V	0	36	0	25	0	29	
t_w Pulse duration	$\overline{PRE}$ or $\overline{CLR}$ low	2 V	100		150		125		
		4.5 V	20		30		25		
		6 V	17		25		21		ns
	CLK high or low	2 V	80		120		100		
		4.5 V	16		24		20		
		6 V	14		20		17		
t_{su} Setup time before CLK↑	Data	2 V	100		150		125		
		4.5 V	20		30		25		
		6 V	17		25		21		ns
	$\overline{PRE}$ or $\overline{CLR}$ inactive	2 V	25		40		30		
		4.5 V	5		8		6		
		6 V	4		7		5		
t_h Hold time data after CLK↑		2 V	0		0		0		
		4.5 V	0		0		0		ns
		6 V	0		0		0		

switching characteristics over recommended operating free-air temperature range (unless otherwise noted), $C_L = 50$ pF (see Note 1)

PARAMETER	FROM (INPUT)	TO (OUTPUT)	V_{CC}	$T_A = 25°C$			SN54HC74		SN74HC74		UNIT
				MIN	TYP	MAX	MIN	MAX	MIN	MAX	
f_{max}			2 V	6	10		4.2		5		
			4.5 V	31	50		21		25		MHz
			6 V	36	60		25		29		
t_{pd}	$\overline{PRE}$ or $\overline{CLR}$	Q or $\overline{Q}$	2 V		70	230		345		290	
			4.5 V		20	46		69		58	
			6 V		15	39		59		49	ns
	CLK	Q or Q	2 V		70	175		250		220	
			4.5 V		20	35		50		44	
			6 V		15	30		42		37	
t_t		Q or $\overline{Q}$	2 V		28	75		110		95	
			4.5 V		8	15		22		19	ns
			6 V		6	13		19		16	

C_{pd}	Power dissipation capacitance per flip-flop	No load, $T_A = 25°C$	35 pF typ

NOTE 1: Load circuit and voltage waveforms are shown in Section 1.

TEXAS
INSTRUMENTS
POST OFFICE BOX 655012 • DALLAS, TEXAS 75265

SN54HC109, SN74HC109
DUAL J-$\overline{K}$ POSITIVE-EDGE-TRIGGERED
FLIP-FLOPS WITH CLEAR AND PRESET

absolute maximum ratings over operating free-air temperature range[†]

Supply voltage, V_{CC} .... -0.5 V to 7 V
Input clamp current, I_{IK} ($V_I < 0$ or $V_I > V_{CC}$) .... ± 20 mA
Output clamp current, I_{OK} ($V_O < 0$ or $V_O > V_{CC}$) .... ± 20 mA
Continuous output current, I_O ($V_O = 0$ to V_{CC}) .... ± 25 mA
Continuous current through V_{CC} or GND pins .... ± 50 mA
Lead temperature 1,6 mm (1/16 in) from case for 60 s: FK or J package .... 300 °C
Lead temperature 1,6 mm (1/16 in) from case for 10 s: D or N package .... 260 °C
Storage temperature range .... -65 °C to 150 °C

[†] Stresses beyond those listed under "absolute maximum ratings" may cause permanent damage to the device. These are stress ratings only, and functional operation of the device at these or any other conditions beyond those indicated under "recommended operating conditions" is not implied. Exposure to absolute-maximum-rated conditions for extended periods may affect device reliability.

recommended operating conditions

			SN54HC109			SN74HC109			UNIT
			MIN	NOM	MAX	MIN	NOM	MAX	
V_{CC}	Supply voltage		2	5	6	2	5	6	V
V_{IH}	High-level input voltage	$V_{CC} = 2$ V	1.5			1.5			V
		$V_{CC} = 4.5$ V	3.15			3.15			
		$V_{CC} = 6$ V	4.2			4.2			
V_{IL}	Low-level input voltage	$V_{CC} = 2$ V	0		0.3	0		0.3	V
		$V_{CC} = 4.5$ V	0		0.9	0		0.9	
		$V_{CC} = 6$ V	0		1.2	0		1.2	
V_I	Input voltage		0		V_{CC}	0		V_{CC}	V
V_O	Output voltage		0		V_{CC}	0		V_{CC}	V
t_t	Input transition (rise and fall) times	$V_{CC} = 2$ V	0		1000	0		1000	ns
		$V_{CC} = 4.5$ V	0		500	0		500	
		$V_{CC} = 6$ V	0		400	0		400	
T_A	Operating free-air temperature		-55		125	-40		85	°C

electrical characteristics over recommended operating free-air temperature range (unless otherwise noted)

PARAMETER	TEST CONDITIONS	V_{CC}	$T_A = 25$ °C			SN54HC109		SN74HC109		UNIT
			MIN	TYP	MAX	MIN	MAX	MIN	MAX	
V_{OH}	$V_I = V_{IH}$ or V_{IL}, $I_{OH} = -20$ μA	2 V	1.9	1.998		1.9		1.9		V
		4.5 V	4.4	4.499		4.4		4.4		
		6 V	5.9	5.999		5.9		5.9		
	$V_I = V_{IH}$ or V_{IL}, $I_{OH} = -4$ mA	4.5 V	3.98	4.30		3.7		3.84		
	$V_I = V_{IH}$ or V_{IL}, $I_{OH} = -5.2$ mA	6 V	5.48	5.80		5.2		5.34		
V_{OL}	$V_I = V_{IH}$ or V_{IL}, $I_{OL} = 20$ μA	2 V		0.002	0.1		0.1		0.1	V
		4.5 V		0.001	0.1		0.1		0.1	
		6 V		0.001	0.1		0.1		0.1	
	$V_I = V_{IH}$ or V_{IL}, $I_{OL} = 4$ mA	4.5 V		0.17	0.26		0.4		0.33	
	$V_I = V_{IH}$ or V_{IL}, $I_{OL} = 5.2$ mA	6 V		0.15	0.26		0.4		0.33	
I_I	$V_I = V_{CC}$ or 0	6 V		± 0.1	± 100		± 1000		± 1000	nA
I_{CC}	$V_I = V_{CC}$ or 0, $I_O = 0$	6 V			4		80		40	μA
C_I		2 to 6 V		3	10		10		10	pF

TEXAS
INSTRUMENTS

POST OFFICE BOX 655012 • DALLAS, TEXAS 75265

SN54HC109, SN74HC109
DUAL J-$\overline{K}$ POSITIVE-EDGE-TRIGGERED
FLIP-FLOPS WITH CLEAR AND PRESET

switching characteristics over recommended operating free-air temperature range (unless otherwise noted), $C_L = 50$ pF (see Note 1)

PARAMETER	FROM (INPUT)	TO (OUTPUT)	V_{CC}	$T_A = 25\,°C$			SN54HC109		SN74HC109		UNIT
				MIN	TYP	MAX	MIN	MAX	MIN	MAX	
f_{max}			2 V	6	10		4.2		5		MHz
			4.5 V	31	50		21		25		
			6 V	36	60		25		29		
t_{pd}	$\overline{PRE}$ or $\overline{CLR}$	Q or $\overline{Q}$	2 V		60	230		345		290	ns
			4.5 V		15	46		69		58	
			6 V		12	39		59		49	
t_{pd}	CLK	Q or $\overline{Q}$	2 V		50	175		250		220	ns
			4.5 V		15	35		50		44	
			6 V		12	30		42		37	
t_t		Q or $\overline{Q}$	2 V		28	75		110		95	ns
			4.5 V		8	15		22		19	
			6 V		6	13		19		16	

C_{pd}	Power dissipation capacitance per flip-flop	No load, $T_A = 25\,°C$	35 pF typ

NOTE 1: Load circuit and voltage waveforms are shown in Section 1.

timing requirements over recommended operating free-air temperature range (unless otherwise noted)

			V_{CC}	$T_A = 25\,°C$		SN54HC109		SN74HC109		UNIT
				MIN	MAX	MIN	MAX	MIN	MAX	
f_{clock}	Clock frequency		2 V	0	6	0	4.2	0	5	MHz
			4.5 V	0	31	0	21	0	25	
			6 V	0	36	0	25	0	29	
t_w	Pulse duration	$\overline{PRE}$ or $\overline{CLR}$ low	2 V	100		150		125		ns
			4.5 V	20		30		25		
			6 V	17		25		21		
		CLK high or low	2 V	80		120		100		
			4.5 V	16		24		20		
			6 V	14		20		17		
t_{su}	Setup time before CLK↑	Data (J, K)	2 V	100		150		125		ns
			4.5 V	20		30		25		
			6 V	17		25		21		
		$\overline{PRE}$ or $\overline{CLR}$ inactive	2 V	25		40		30		
			4.5 V	5		8		6		
			6 V	4		7		5		
t_h	Hold time, data after CLK↑		2 V	0		0		0		ns
			4.5 V	0		0		0		
			6 V	0		0		0		

TEXAS
INSTRUMENTS
POST OFFICE BOX 655012 • DALLAS, TEXAS 75265

SN5446A, '47A,
SN7446A, '47A,
BCD-TO-SEVEN-SEGMENT DECODERS/DRIVERS

TYPE	DRIVER OUTPUTS				TYPICAL	PACKAGES
	ACTIVE LEVEL	OUTPUT CONFIGURATION	SINK CURRENT	MAX VOLTAGE	POWER DISSIPATION	
SN5446A	low	open-collector	40 mA	30 V	320 mW	J, W
SN5447A	low	open-collector	40 mA	15 V	320 mW	J, W
SN7446A	low	open-collector	40 mA	30 V	320 mW	J, N
SN7447A	low	open-collector	40 mA	15 V	320 mW	J, N

SEGMENT IDENTIFICATION

$$\begin{array}{cccccccccccccccc} 0 & 1 & 2 & 3 & 4 & 5 & 6 & 7 & 8 & 9 & 10 & 11 & 12 & 13 & 14 & 15 \end{array}$$

NUMERICAL DESIGNATIONS AND RESULTANT DISPLAYS

TTL Devices

'46A, '47A, 'LS47 FUNCTION TABLE (T1)

DECIMAL OR FUNCTION	INPUTS						BI/RBO†	OUTPUTS							NOTE
	LT	RBI	D	C	B	A		a	b	c	d	e	f	g	
0	H	H	L	L	L	L	H	ON	ON	ON	ON	ON	ON	OFF	
1	H	X	L	L	L	H	H	OFF	ON	ON	OFF	OFF	OFF	OFF	
2	H	X	L	L	H	L	H	ON	ON	OFF	ON	ON	OFF	ON	
3	H	X	L	L	H	H	H	ON	ON	ON	ON	OFF	OFF	ON	
4	H	X	L	H	L	L	H	OFF	ON	ON	OFF	OFF	ON	ON	
5	H	X	L	H	L	H	H	ON	OFF	ON	ON	OFF	ON	ON	
6	H	X	L	H	H	L	H	OFF	OFF	ON	ON	ON	ON	ON	
7	H	X	L	H	H	H	H	ON	ON	ON	OFF	OFF	OFF	OFF	1
8	H	X	H	L	L	L	H	ON	ON	ON	ON	ON	ON	ON	
9	H	X	H	L	L	H	H	ON	ON	ON	OFF	OFF	ON	ON	
10	H	X	H	L	H	L	H	OFF	OFF	OFF	ON	ON	OFF	ON	
11	H	X	H	L	H	H	H	OFF	OFF	ON	ON	OFF	OFF	ON	
12	H	X	H	H	L	L	H	OFF	ON	OFF	OFF	OFF	ON	ON	
13	H	X	H	H	L	H	H	ON	OFF	OFF	ON	OFF	ON	ON	
14	H	X	H	H	H	L	H	OFF	OFF	OFF	ON	ON	ON	ON	
15	H	X	H	H	H	H	H	OFF	OFF	OFF	OFF	OFF	OFF	OFF	
BI	X	X	X	X	X	X	L	OFF	OFF	OFF	OFF	OFF	OFF	OFF	2
RBI	H	L	L	L	L	L	L	OFF	OFF	OFF	OFF	OFF	OFF	OFF	3
LT	L	X	X	X	X	X	H	ON	ON	ON	ON	ON	ON	ON	4

H = high level, L = low level, X = irrelevant

NOTES: 1. The blanking input ($\overline{BI}$) must be open or held at a high logic level when output functions 0 through 15 are desired. The ripple-blanking input ($\overline{RBI}$) must be open or high if blanking of a decimal zero is not desired.
2. When a low logic level is applied directly to the blanking input ($\overline{BI}$), all segment outputs are off regardless of the level of any other input.
3. When ripple-blanking input ($\overline{RBI}$) and inputs A, B, C, and D are at a low level with the lamp test input high, all segment outputs go off and the ripple-blanking output ($\overline{RBO}$) goes to a low level (response condition).
4. When the blanking input/ripple blanking output ($\overline{BI}/\overline{RBO}$) is open or held high and a low is applied to the lamp-test input, all segment outputs are on.

†$\overline{BI}/\overline{RBO}$ is wire AND logic serving as blanking input ($\overline{BI}$) and/or ripple-blanking output ($\overline{RBO}$).

TEXAS INSTRUMENTS
POST OFFICE BOX 655012 • DALLAS, TEXAS 75265

Reprinted by permission of Texas Instruments.

SN5446A, SN5447A, SN7446A, SN7447A
BCD-TO-SEVEN-SEGMENT DECODERS/DRIVERS

absolute maximum ratings over operating free-air temperature range (unless otherwise noted)

Supply voltage, V_{CC} (see Note 1)	7 V
Input voltage	5.5 V
Current forced into any output in the off state	1 mA
Operating free-air temperature range: SN5446A, SN5447A	$-55°C$ to $125°C$
SN7446A, SN7447A	$0°C$ to $70°C$
Storage temperature range	$-65°C$ to $150°C$

NOTE 1: Voltage values are with respect to network ground terminal.

recommended operating conditions

		SN5446A			SN5447A			SN7446A			SN7447A			UNIT
		MIN	NOM	MAX	MIN	NOM	MAX	MIN	NOM	MAX	MIN	NOM	MAX	
Supply voltage, V_{CC}		4.5	5	5.5	4.5	5	5.5	4.75	5	5.25	4.75	5	5.25	V
Off-state output voltage, $V_{O(off)}$	a thru g			30			15			30			15	V
On-state output current, $I_{O(on)}$	a thru g			40			40			40			40	mA
High-level output current, I_{OH}	$\overline{BI}/\overline{RBO}$			-200			-200			-200			-200	μA
Low-level output current, I_{OL}	$\overline{BI}/\overline{RBO}$			8			8			8			8	mA
Operating free-air temperature, T_A		-55		125	-55		125	0		70	0		70	°C

electrical characteristics over recommended operating free-air temperature range (unless otherwise noted)

	PARAMETER		TEST CONDITIONS†		MIN	TYP‡	MAX	UNIT
V_{IH}	High-level input voltage				2			V
V_{IL}	Low-level input voltage						0.8	V
V_{IK}	Input clamp voltage		V_{CC} = MIN, $I_I = -12$ mA				-1.5	V
V_{OH}	High-level output voltage	$\overline{BI}/\overline{RBO}$	V_{CC} = MIN, V_{IH} = 2 V, V_{IL} = 0.8 V, $I_{OH} = -200\ \mu A$		2.4	3.7		V
V_{OL}	Low-level output voltage	$\overline{BI}/\overline{RBO}$	V_{CC} = MIN, V_{IH} = 2 V, V_{IL} = 0.8 V, I_{OL} = 8 mA			0.27	0.4	V
$I_{O(off)}$	Off-state output current	a thru g	V_{CC} = MAX, V_{IH} = 2 V, V_{IL} = 0.8 V, $V_{O(off)}$ = MAX				250	μA
$V_{O(on)}$	On-state output voltage	a thru g	V_{CC} = MIN, V_{IH} = 2 V, V_{IL} = 0.8 V, $I_{O(on)}$ = 40 mA			0.3	0.4	V
I_I	Input current at maximum input voltage	Any input except $\overline{BI}/\overline{RBO}$	V_{CC} = MAX, V_I = 5.5 V				1	mA
I_{IH}	High-level input current	Any input except $\overline{BI}/\overline{RBO}$	V_{CC} = MAX, V_I = 2.4 V				40	μA
I_{IL}	Low-level input current	Any input except $\overline{BI}/\overline{RBO}$	V_{CC} = MAX, V_I = 0.4 V				-1.6	mA
		$\overline{BI}/\overline{RBO}$					-4	
I_{OS}	Short-circuit output current	$\overline{BI}/\overline{RBO}$	V_{CC} = MAX				-4	mA
I_{CC}	Supply current		V_{CC} = MAX, See Note 2	SN54'		64	85	mA
				SN74'		64	103	

†For conditions shown as MIN or MAX, use the appropriate value specified under recommended operating conditions.
‡All typical values are at V_{CC} = 5 V, T_A = 25°C.
NOTE 2: I_{CC} is measured with all outputs open and all inputs at 4.5 V.

switching characteristics, V_{CC} = 5 V, T_A = 25°C

	PARAMETER	TEST CONDITIONS	MIN	TYP	MAX	UNIT
t_{off}	Turn-off time from A input	C_L = 15 pF, R_L = 120 Ω, See Note 3			100	ns
t_{on}	Turn-on time from A input				100	
t_{off}	Turn-off time from $\overline{RBI}$ input				100	ns
t_{on}	Turn-on time from $\overline{RBI}$ input				100	

NOTE 3: Load circuits and voltage waveforms are shown in Section 1.

TEXAS INSTRUMENTS
POST OFFICE BOX 655012 • DALLAS, TEXAS 75265

TTL Devices

LED Displays

7 SEGMENT, SUPER BRIGHT

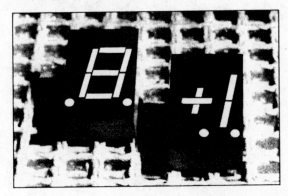

FEATURES

- 0.3″, 0.43″ and 0.56″ character heights
- Available in Super Bright RED
- Application: Numerical Readout for Instrument and Industrial products
- Industry pin for pin compatibility
- Both common cathode or common anode are available

	Number of Digits	Common		Color		Number of Pins
		Cathode	Anode	Display	Face	
.3 inch	1	AND332S		Red	Red	10
.3 inch	1		AND333S/AND335S*	Red	Red	14
.3 inch	1/2	AND334S		Red	Red	14
.43 inch	1	AND342S	AND343S/AND345S*	Red	Red	14
.43 inch	1/2	AND344S		Red	Red	14
.56 inch	1	AND362S	AND363S	Red	Red	10
.56 inch	1/2	AND364S	AND365S	Red	Gray	10

Absolute Maximum Ratings (T = 25°C)

Characteristic	Symbol	Rating	Unit
DC Forward Current/Segment	$I_F(DC)/SEG$	20	mA
Pulse Forward Current/Segment	I_{FP}/SEG	110	mA
Reverse Voltage/Segment	V_R	6	V
Operating Temperature Range	T_{opr}	40 to +85	°C
Storage Temperature Range	T_{stg}	40 to +85	°C

Electro-Optical Characteristics

Characteristics	Symbol	Condition	Min.	Typ.	Max.	Unit
Forward Voltage	V_F	$I_F = 10mA$	1.7	2	2.5	V_F
Reverse Current	I_R	$V_R = 6V$			5	μA
Luminous Intensity Per Segment						
AND33XS	I_V	$I_F = 10mA$	0.9	1.85		mcd
AND34XS	I_V	$I_F = 10mA$	0.6	1.28		mcd
AND36XS	I_V	$I_F = 10mA$	1.0	2.00		mcd
Luminous Intensity Matching Ratio	I_V-M	$I_F = 10mA$			2.3	
Peak Emission Wave Length	λ_P	$I_F = 10mA$		635		nm
Spectral Line Half Width	$\Delta\lambda$	$I_F = 10mA$		40		nm

* AND333/343 Series, right hand decimal point
  AND335/345 Series, left hand decimal point

Reliability Data

The following reliability tables and models have been abstracted from Military Handbook for Reliability Prediction of Electronic Systems, MIL-HDBK 217E, Published by the United States Department of Defense, 1986.

Table R.1 Codes are used for specifying the environmental application for each device, π_E.

Code	Application
G_B – Ground, Benign	Laboratory Use
G_F – Ground, Fixed	Installation in fixed racks
G_M – Ground, Mobile	Installed on wheeled or tracked vehicles
S_F – Space, Flight	Earth orbit
M_P – Manpack	Transported manually while in use
A_{IC} – Airborne, Inhabited, Cargo	Cargo compartments occupied by aircrew
A_{UC} – Airborne, Uninhabited, Cargo	Cargo compartments, bomb & equipment bays
A_{RW} – Airborne, Rotary Winged	Equipment installed on helicopters
M_L – Missile Launch	Conditions related to missile launch

The following factors are defined as follows for each device listed in this section:

λ_P, General model for failure rate per component.

SWITCHES:

λ_b, Base failure rate,

Π_E, Environmental mode factor,

Π_Q, Quality factor,

Π_C, Contact form factor / Configuration factor,

Π_{CYC}, Cycling rate factor,

Π_L, Load stress factor,

Π_U, Power on-off switch factor.

FUSES:

λ_b, Base failure rate,

Π_E, Environmental mode factor.

RESISTORS and POTENTIOMETERS:

λ_b, Base failure rate,

Π_E, Environmental mode factor,

Π_Q, Quality factor,

Π_R, Resistance mode factor,

Π_{TAPS}, Potentiometer taps factor,

Π_V, Voltage factor.

CAPACITORS:

λ_b, Base failure rate,

Π_E, Environmental mode factor,

Π_Q, Quality factor,

Π_{CV}, Capacitance factor,

Π_C, Construction factor.

INDUCTORS and TRANSFORMERS:

λ_b, Base failure rate,

Π_E, Environmental mode factor,

Π_Q, Quality factor,

Π_C, Construction factor.

TRANSISTORS:

λ_b, Base failure rate,

Π_E, Environmental mode factor,

Π_Q, Quality factor,

Π_A, Applications factor,

Π_R, Power rating factor,

Π_C, Complexity factor,

Π_{S2}, Voltage stress factor.

DIODES:

λ_b, Base failure rate,

Π_E, Environmental mode factor,

Π_Q, Quality factor,

Π_A, Applications factor,

Π_R, Current rating factor,

Π_C, Construction factor,

Π_{S2}, Voltage stress factor.

OPTO-ELECTRONIC DEVICES:

λ_b, Base failure rate,

Π_E, Environmental mode factor,

Π_Q, Quality factor,

Π_T, Temperature factor.

MICROELECTRONIC DEVICES:

Π_E, Environmental mode factor,

Π_Q, Quality factor,

Π_T, Temperature acceleration factor,

Π_V, Volatge stress derating factor,

Π_L, Device learning factor,

C1, Circuit complexity factor,

C2, Package complexity factor.

SWITCHES

Toggle or Pushbutton:
The general model for these switches is as follows:

$$\lambda_P = \lambda_b (\Pi_E \cdot \Pi_C \cdot \Pi_{CYC} \cdot \Pi_L) \text{ failures}/10^6 \text{ hours.}$$

Table R.2 Basic Failure Rates

Description	λ_b	
	MIL-Spec	Industrial
Snap Action	0.00045	0.034
Non-Snap Action	0.0027	0.04

Table R.3 Environmental Factors

Environment	Π_E
G_B	1.0
G_F	2.9
G_M	14.0
S_F	1.0
M_P	21.0
A_{IC}	8.0
A_{UC}	10.0
A_{RW}	46.0
M_L	71.0

Table R.4 Contact Form Factors

Contact Form	Π_C
SPST	1.0
DPST	1.5
SPDT	1.75
3PST	2.0
4PST	2.5
DPDT	3.0
3PDT	4.25
4PDT	5.5
6PDT	8.0

Table R.5 Cycling Rate Factors

Switching Cycles per Hour	Π_{CYC}
<= 1 cycle/hour	1.0
> 1 cycle/hour	# of cycles per hour

Table R.6 Load Stress Factor

$$S = \frac{\text{operating load current}}{\text{rated resistive load current}}$$

Load Type	Π_L
Resistive	$e^{(S/0.8)}$
Inductive	$e^{(S/0.4)}$
Lamp	$e^{(S/0.2)}$

FUSES

The general model for fuses is as follows:

$$\lambda_P = \lambda_b \cdot \Pi_E \text{ failures/10}^6 \text{ hours.}$$

$$\lambda_b = 0.010$$

Table R.7 Environmental Factors

Environment	Π_E
G_B	1.0
G_F	2.3
G_M	7.5
S_F	1.8
M_P	8.1
A_{IC}	6.8
A_{UC}	13.0
A_{RW}	16.0
M_L	22.0

RESISTORS

The general model for a resistor is as follows:

$$\lambda_P = \lambda_b (\Pi_E \cdot \Pi_R \cdot \Pi_Q) \text{ failures/10}^6 \text{ hours}$$

The general model for a variable resistor is as follows:

$$\lambda_P = \lambda_b (\Pi_{TAPS} \cdot \Pi_R \cdot \Pi_V \cdot \Pi_Q \cdot \Pi_E) \text{ failures/10}^6 \text{ hours.}$$

S is the electrical stress and is the ratio of the operating power to the rated power, except for potentiometers.

Table R.8 Environmental Factors

Environment	Π_E			
	Carbon Comp.	Wire Wound	Pots.	Metal Film
G_B	1.0	1.0	1.0	1.0
G_F	2.9	1.5	1.8	2.4
G_M	8.3	8.3	17.0	7.8
M_P	8.5	11.0	21.0	8.8
S_F	1.0	0.6	1.0	0.4
A_{IC}	3.0	1.5	30.0	2.5
A_{UC}	5.0	5.6	40.0	7.0
A_{RW}	19.0	23.0	46.0	19.0
M_L	29.0	36.0	71.0	30.0

Table R.9 Resistance Factors

Resistance Range	Π_R			
	Carbon Comp.	Wire Wound	Pots.	Metal Film
Up to 10K	1.0	1.0	1.0	1.0
>10K–50K	1.0	–	1.0	1.0
>50K–100K	1.0	–	1.1	1.0
>100K–200K	1.1	–	1.2	1.1
>200K–500K	1.1	–	1.4	1.1
>500K–1M	1.1	–	1.8	1.1
>1M–10M	1.6	–	–	1.6
>10M	2.5	–	–	2.5

Table R.10 Quality Factors

Failure Rate Level	Π_Q			
	Carbon Comp.	Wire Wound	Pots.	Metal Film
MIL-Spec	5.0	5.0	2.5	5.0
Industrial	15.0	15.0	5.0	15.0

POTENTIOMETERS

Table R.11 Voltage Factors For Potentiometers

Ratio of Applied Voltage to Rated Voltage	Π_V
0.0 to 0.8	1.00
> 0.8 to 0.9	1.05
> 0.9 to 1.0	1.20

$$V_{applied} = (R \cdot P_{applied})^{1/2}$$

$$S = \frac{P_{applied}}{\Pi_{eff} \cdot P_{rated}}$$

$P_{applied} = (V_{in}^2)/R_P$; when wiper lead is disconnected.

Π_{eff} = correction factor.

P_{rated} = power rating of the potentiometer.

$$\Pi_{eff} = \frac{R_L^2}{R_L^2 + K_H \cdot (R_P^2 + 2 \cdot R_P \cdot R_L)}$$

R_L = Load resistance, total resistance between wiper arm and one end of the potentiometer (lowest value).

R_P = Total potentiometer resistance.

$K_H = 0.5$

$$\Pi_{taps} = (N_{taps}^{3/2}) / 25 + 0.792$$

N_{taps} = number of potentiometers taps, including the wiper and the end terminals.

Table R.12   Carbon Composition Resistors, Base Failure Rates, λ_b

S, RATIO OF OPERATING TO RATED POWER

TEMP ($^\circ$C)	.1	.2	.3	.4	.5	.6	.7	.8	.9	1.0
0	.00007	.00009	.00010	.00012	.00015	.00017	.00020	.00024	.00028	.00033
10	.00011	.00013	.00015	.00018	.00021	.00025	.00030	.00036	.00043	.00051
20	.00015	.00018	.00022	.00026	.00031	.00037	.00045	.00053	.00064	.00076
30	.00022	.00026	.00031	.00038	.00046	.00055	.00066	.00079	.00096	.0011
40	.00031	.00038	.00045	.00055	.00067	.00081	.00098	.0012	.0014	.0017
50	.00044	.00054	.00066	.00080	.00098	.0012	.0014	.0018	.0021	.0026
60	.00063	.00078	.00095	.0012	.0014	.0017	.0021	.0026	.0032	.0039
70	.00090	.0011	.0014	.0017	.0021	.0026	.0032	.0039	.0048	.0059
80	.0013	.0016	.0020	.0025	.0031	.0038	.0047	.0058		
90	.0018	.0023	.0029	.0036	.0045	.0056				
100	.0026	.0033	.0041	.0052	.0065					
110	.0038	.0047	.0060							
120	.0054									

Table R.13   Potentiometer, Base Failure Rates, λ_b

S, RATIO OF OPERATING TO RATED POWER

TEMP ($^\circ$C)	.1	.2	.3	.4	.5	.6	.7	.8	.9	1.0
0	.027	.028	.030	.031	.032	.034	.035	.037	.038	.040
10	.028	.029	.031	.033	.034	.036	.038	.040	.042	.045
20	.029	.031	.033	.035	.037	.040	.042	.045	.048	.051
30	.031	.033	.036	.038	.041	.045	.048	.052	.056	.060
40	.033	.036	.039	.043	.047	.051	.056	.061	.067	.073
50	.036	.039	.044	.049	.054	.060	.067	.074	.082	.091
60	.039	.045	.050	.057	.065	.073	.083	.094	.11	.12
70	.045	.052	.060	.069	.080	.093	.11	.12	.14	.17
80	.053	.063	.074	.088	.10	.12	.15	.17		
90	.065	.079	.096	.12	.14	.17				
100	.084	.11	.13	.16						
110	.11	.15								

Table R.14 Wire-Wound Resistors, Base Failure Rates, λ_b

TEMP ($^\circ$C)	S, RATIO OF OPERATING TO RATED POWER									
	.1	.2	.3	.4	.5	.6	.7	.8	.9	1.0
0	.0042	.0051	.0062	.0076	.0093	.011	.014	.017	.021	.025
10	.0045	.0055	.0068	.0084	.010	.013	.016	.019	.024	.029
20	.0048	.0060	.0074	.0092	.011	.014	.017	.022	.027	.033
30	.0052	.0065	.0081	.010	.013	.016	.020	.025	.031	
40	.0056	.0071	.0089	.011	.014	.018	.022	.028	.035	
50	.0061	.0077	.0097	.012	.016	.020	.025	.032	.040	
60	.0066	.0084	.011	.014	.017	.022	.028	.036		
70	.0072	.0092	.012	.015	.020	.025	.032	.042		
80	.0078	.010	.013	.017	.022	.028	.037	.048		
90	.0085	.011	.014	.019	.025	.032	.042	.055		
100	.0093	.012	.016	.021	.028	.037	.048			
110	.010	.014	.018	.024	.031	.042	.055			
120	.011	.015	.020	.027	.036	.047	.063			
130	.012	.017	.022	.030	.040	.054				
140	.014	.019	.025	.034	.046	.062				
150	.015	.021	.028	.038	.052	.071				
160	.017	.023	.032	.043	.060					
170	.019	.026	.036	.049	.068					
180	.021	.029	.040	.056	.078					
190	.023	.033	.046	.064						
200	.026	.037	.052	.074						
210	.029	.042	.059	.084						
220	.033	.047	.068	.097						
230	.037	.053	.077							
240	.042	.061	.088							
250	.047	.069	.10							
260	.054	.079								
270	.061	.091								

Table R.15   Metal Film Resistors, Base Failure Rates, λ_b

S, RATIO OF OPERATING TO RATED POWER

TEMP ($^{\circ}$C)	.1	.2	.3	.4	.5	.6	.7	.8	.9	1.0
0	.00061	.00067	.00074	.00082	.00091	.0010	.0011	.0012	.0014	.0015
10	.00067	.00074	.00082	.00091	.0010	.0011	.0012	.0014	.0015	.0017
20	.00073	.00082	.00091	.0010	.0011	.0013	.0014	.0016	.0017	.0019
30	.00080	.00090	.0010	.0011	.0013	.0014	.0016	.0017	.0019	.0022
40	.00088	.00099	.0011	.0012	.0014	.0016	.0017	.0020	.0022	.0025
50	.00096	.0011	.0012	.0014	.0015	.0017	.0020	.0022	.0025	.0028
60	.0011	.0012	.0013	.0015	.0017	.0019	.0022	.0025	.0028	.0032
70	.0012	.0013	.0015	.0017	.0019	.0022	.0025	.0028	.0032	.0036
80	.0013	.0014	.0016	.0019	.0021	.0024	.0028	.0031	.0036	.0041
90	.0014	.0016	.0018	.0021	.0024	.0027	.0031	.0035	.0040	.0046
100	.0015	.0017	.0020	.0023	.0026	.0030	.0035	.0040	.0045	.0052
110	.0017	.0019	.0022	.0025	.0029	.0034	.0039	.0045	.0051	.0059
120	.0018	.0021	.0024	.0028	.0033	.0038	.0043	.0050	.0058	.0067
130	.0020	.0023	.0027	.0031	.0036	.0042	.0049	.0056	.0065	
140	.0022	.0026	.0030	.0035	.0040	.0047	.0054			
150	.0024	.0028	.0033	.0038	.0045					
160	.0026	.0031	.0036							
170	.0029									

CAPACITORS

The general model for a capacitor is as follows:

$$\lambda_P = \lambda_b \ (\Pi_E \cdot \Pi_{CV} \cdot \Pi_Q \cdot \Pi_C) \text{ failures}/10^6 \text{ hours.}$$

Table R.16   Environmental Factors

Environment	Π_E			
	Al. Elect.	Mylar	Mica	Ceramic Disc
G_B	1.0	1.0	1.0	1.0
G_F	2.4	1.9	2.4	1.6
G_M	12.0	9.3	8.8	7.8
S_F	1.0	1.0	1.0	0.8
M_P	12.0	11.0	11.0	11.0
A_{IC}	9.5	3.5	3.5	3.0
A_{UC}	25.0	10.0	15.0	7.5
A_{RW}	27.0	23.0	23.0	24.0
M_L	41.0	36.0	36.0	36.0

Table R.17 Capacitance Factors

	Al. Elect.	Mylar	Mica	Ceramic Disc
Π_{CV}	$0.32 \cdot C^{0.19}$ C is in μF	$1.1 \cdot C^{0.085}$ C is in μF	$0.45 \cdot C^{0.14}$ C is in pF	$0.41 \cdot C^{0.11}$ C is in pF

Table R.18 Quality Factors

Failure Rate Level	Π_Q			
	Al. Elect.	Mylar	Mica	Ceramic Disc
MIL-Spec	3.0	3.0	3.0	3.0
Industrial	10.0	10.0	15.0	10.0

$\Pi_C = 1.0$: Construction Factor for all capacitors listed.

Table R.19 Aluminum Electrolytic Capacitors, Base Failure Rates, λ_b

S, RATIO OF OPERATING TO RATED VOLTAGE

TEMP ($^\circ$C)	.1	.2	.3	.4	.5	.6	.7	.8	.9	1.0
0	.0064	.0067	.0074	.0089	.011	.015	.020	.026	.034	.045
10	.0078	.0082	.0090	.011	.014	.018	.024	.032	.042	.055
20	.0099	.010	.011	.014	.017	.023	.030	.040	.053	.069
30	.013	.014	.015	.018	.023	.030	.040	.053	.070	.091
40	.018	.019	.021	.025	.031	.041	.055	.073	.096	.13
50	.026	.027	.030	.036	.046	.060	.080	.11	.14	.18
60	.041	.042	.047	.056	.071	.093	.12	.16	.22	.28
70	.068	.070	.078	.093	.12	.15	.21	.27	.36	.47
80	.12	.13	.14	.17	.21	.28	.37	.49	.65	.85

Table R.20 Ceramic Capacitors, Base Failure Rates, λ_b

S, RATIO OF OPERATING TO RATED VOLTAGE

TEMP ($^\circ$C)	.1	.2	.3	.4	.5	.6	.7	.8	.9	1.0
0	.00062	.00077	.0012	.0020	.0034	.0054	.0082	.012	.017	.023
10	.00063	.00079	.0012	.0021	.0034	.0055	.0084	.012	.017	.023
20	.00065	.00081	.0013	.0021	.0035	.0056	.0086	.013	.018	.024
30	.00067	.00083	.0013	.0022	.0036	.0058	.0088	.013	.018	.024
40	.00068	.00085	.0013	.0022	.0037	.0059	.0090	.013	.018	.025
50	.00070	.00088	.0014	.0023	.0038	.0061	.0093	.013	.019	.026
60	.00072	.00090	.0014	.0023	.0039	.0062	.0095	.014	.019	.026
70	.00074	.00092	.0014	.0024	.0040	.0064	.0097	.014	.020	.027
80	.00076	.00094	.0015	.0025	.0041	.0066	.010	.015	.020	.028
90	.00077	.00097	.0015	.0025	.0042	.0067	.010	.015	.021	.028
100	.00079	.00099	.0015	.0026	.0043	.0069	.010	.015	.021	.029
110	.00081	.0010	.0016	.0026	.0044	.0071	.011	.016	.022	.030
120	.00084	.0010	.0016	.0027	.0045	.0072	.011	.016	.023	.031

Table R.21 Mylar Capacitors, Base Failure Rates, λ_b

S, RATIO OF OPERATING TO RATED VOLTAGE

TEMP ($^\circ$C)	.1	.2	.3	.4	.5	.6	.7	.8	.9	1.0
0	.00099	.0010	.0012	.0020	.0040	.0085	.017	.033	.058	.098
10	.0010	.0010	.0012	.0020	.0040	.0086	.017	.033	.058	.098
20	.0010	.0010	.0012	.0020	.0041	.0086	.017	.033	.059	.099
30	.0010	.0010	.0012	.0020	.0041	.0087	.018	.033	.059	.099
40	.0010	.0011	.0013	.0020	.0041	.0088	.018	.034	.060	.10
50	.0011	.0011	.0013	.0021	.0043	.0090	.018	.035	.062	.10
60	.0011	.0011	.0014	.0022	.0044	.0094	.019	.036	.064	.11
70	.0012	.0012	.0015	.0024	.0048	.010	.020	.039	.069	.12
80	.0013	.0014	.0016	.0026	.0054	.011	.023	.044	.077	.13
90	.0016	.0016	.0020	.0032	.0065	.014	.028	.053	.094	.16
100	.0022	.0022	.0027	.0043	.0087	.019	.038	.071	.13	.21
110	.0035	.0036	.0043	.0069	.014	.030	.060	.11	.20	.34
120	.0073	.0075	.0090	.015	.029	.062	.13	.24	.43	.72

COILS AND TRANSFORMERS

The general model for an inductive component is as follows:

$$\lambda_P = \lambda_b \ (\Pi_E \cdot \Pi_Q \cdot \Pi_C) \ \text{failures}/10^6 \ \text{hours.}$$

Table R.22   Environmental Factors

Environment	Π_E
G_B	1.0
G_F	5.7
G_M	12.0
S_F	1.0
M_P	11.0
A_{IC}	4.5
A_{UC}	6.5
A_{RW}	24.0
M_L	36.0

Table R.23   Quality Factors

Failure Rate Level	Π_Q
MIL–Spec	8.0
Industrial	30.0

$\Pi_C = 1.0$: Construction Factor

$T_{HS} = T_A + 1.1(\Delta T) \ ^\circ C$

T_{HS} : Hot Spot Temperature

T_A : Ambient Temperature

ΔT : Temperature Rise

Table R.24   Inductor and Transformer Base Failure Rates, λ_b

T_{HS}	MAXIMUM RATED OPERATING TEMPERATURE		
	85 $^\circ$C	105 $^\circ$C	130 $^\circ$C
0	0.0017	0.0019	0.0016
5	0.0017	0.0019	0.0017
10	0.0017	0.0019	0.0017
15	0.0018	0.0019	0.0017
20	0.0019	0.0019	0.0018
25	0.0020	0.0020	0.0018
30	0.0021	0.0020	0.0019
35	0.0023	0.0021	0.0010
40	0.0025	0.0022	0.0020
45	0.0029	0.0023	0.0021
50	0.0034	0.0024	0.0022
55	0.0041	0.0026	0.0023
60	0.0053	0.0029	0.0024
65	0.0073	0.0032	0.0026
70	0.0108	0.0036	0.0028
75	0.0175	0.0042	0.0030
80	0.0319	0.0051	0.0033
85	0.0666	0.0064	0.0036
90		0.0084	0.0040
95		0.0116	0.0046
100		0.0171	0.0052
105		0.0271	0.0061
110			0.0072
115			0.0087
120			0.0107
125			0.0134
130			0.0172

TRANSISTORS

The general model for a conventional transistor is as follows:

$$\lambda_P = \lambda_b \left(\Pi_E \cdot \Pi_A \cdot \Pi_Q \cdot \Pi_R \cdot \Pi_{S2} \cdot \Pi_C \right) \text{ failures}/10^6 \text{ hours.}$$

Silicon - NPN and PNP Transistors.

Table R.25 Environmental Factors

Environment	Π_E
G_B	1.0
G_F	5.8
G_M	18.0
S_F	0.4
M_P	12.0
A_{IC}	9.5
A_{UC}	15.0
A_{RW}	27.0
M_L	41.0

Table R.27 Quality Factors

Quality Level	Π_Q
Industrial	6.0
Plastic	12.0

Table R.28 Applications Factors

Application	Π_A
Linear	1.5
Switch	0.7
Si, low noise	15.0

Table R.26 Power Rating Factors

Power Rating (W)	Π_R
<= 1	1.0
> 1 to 5	1.5
> 5 to 20	2.0
> 20 to 50	2.5
> 50 to 200	5.0

$$S_2 = \frac{V_{CE, \text{ applied}}}{V_{CEO, \text{ rated}}} \cdot 100 \qquad \text{For } S_2 \geq 25: \quad \Pi_{S2} = 0.14 \times 10^{(0.0133)S_2}$$

$$\text{For } S_2 < 25: \quad \Pi_{S2} = 0.3$$

$$\Pi_C = 1 \quad \text{For Transistors Listed}$$

Table R.29 NPN Silicon Transistors, Base Failure Rate, λ_b, in Failures Per 10^6 Hours

TEMP ($^\circ$C)	POWER STRESS									
	.1	.2	.3	.4	.5	.6	.7	.8	.9	1.0
0	.00049	.00060	.00071	.00084	.00099	.0012	.0014	.0017	.0021	.0027
10	.00056	.00067	.00079	.00093	.0011	.0013	.0016	.0019	.0025	.0034
20	.00063	.00075	.00089	.0010	.0012	.0015	.0018	.0023	.0030	.0043
30	.00071	.00084	.00099	.0012	.0014	.0017	.0021	.0027	.0038	
40	.00079	.00093	.0011	.0013	.0016	.0019	.0025	.0034	.0049	
50	.00089	.0010	.0012	.0015	.0018	.0023	.0030	.0043		
60	.00099	.0012	.0014	.0017	.0021	.0027	.0038			
70	.0011	.0013	.0016	.0019	.0025	.0034	.0049			
80	.0012	.0015	.0018	.0023	.0030	.0043				
90	.0014	.0017	.0021	.0027	.0038					
100	.0016	.0019	.0025	.0034	.0049					
110	.0018	.0023	.0030	.0043						
120	.0021	.0027	.0038							
130	.0025	.0034	.0049							
140	.0030	.0043								
150	.0038									
160	.0049									

Table R.30 PNP Silicon Transistors, Base Failure Rate, λ_b, in Failures Per 10^6 Hours

TEMP ($^\circ$C)	POWER STRESS									
	.1	.2	.3	.4	.5	.6	.7	.8	.9	1.0
0	.00065	.00082	.0010	.0012	.0015	.0018	.0021	.0026	.0033	.0044
10	.00076	.00095	.0012	.0014	.0017	.0020	.0024	.0030	.0040	.0056
20	.00088	.0011	.0013	.0016	.0019	.0023	.0028	.0036	.0050	.0077
30	.0010	.0012	.0015	.0018	.0021	.0026	.0033	.0044	.0065	
40	.0012	.0014	.0017	.0020	.0024	.0030	.0040	.0056	.0092	
50	.0013	.0016	.0019	.0023	.0028	.0036	.0050	.0077		
60	.0015	.0018	.0021	.0026	.0033	.0044	.0065			
70	.0017	.0020	.0024	.0030	.0040	.0056	.0092			
80	.0019	.0023	.0028	.0036	.0050	.0077				
90	.0021	.0026	.0033	.0044	.0065					
100	.0024	.0030	.0040	.0056	.0092					
110	.0028	.0036	.0050	.0077						
120	.0033	.0044	.0065							
130	.0040	.0056	.0092							
140	.0050	.0077								
150	.0065									
160	.0092									

$$\text{Power Stress} = \frac{\text{Operating Power}}{\text{Rated Power}}$$

DIODES

The general model for a diode is as follows:

$$\lambda_P = \lambda_b (\Pi_E \cdot \Pi_Q \cdot \Pi_R \cdot \Pi_A \cdot \Pi_{S2} \cdot \Pi_C) \text{ failures/}10^6 \text{ hours.}$$

Silicon – General Purpose Diodes

Table R.31   Environmental Factors

Environment	Π_E
G_B	1.0
G_F	3.9
G_M	18.0
S_F	1.0
M_P	12.0
A_{IC}	15.0
A_{UC}	25.0
A_{RW}	27.0
M_L	41.0

Table R.32 Current Rating Factors

Current Rating (A)	Π_R
<= 1	1.0
> 1 to 3	1.5
> 3 to 10	2.0
> 10 to 20	4.0
> 20 to 50	10.0

Table R.33   Quality Factors

Quality Level	Π_Q
Industrial	7.5
Plastic	15.0

Table R.34   Applications Factors

Application	Π_A
Analog Circuits (< 500 mA)	1.0
Switching (< 500 mA)	0.6
Power Rectifier (>= 500 mA)	1.5
Power Rectifier (HV stacks, V_{max} > 600 V)	2.5 / junction

Table R.35   Voltage Stress Factors

$$S_2 = \frac{V_{R, \text{ applied}}}{V_{R, \text{ rated}}} \cdot 100\%$$

where V_R = Diode Reverse Voltage

S_2 (percent)	Π_{S2}
0 to 60	0.70
> 60 to 70	0.75
> 70 to 80	0.80
> 80 to 90	0.90
> 90 to 100	1.0

Π_C = 1.0 : Contruction Factor

Table R.36　Silicon Diodes, Base Failure Rate λ_b, in Failures Per 10^6 Hours

TEMP ($^\circ$C)	CURRENT STRESS, S									
	.1	.2	.3	.4	.5	.6	.7	.8	.9	1.0
0	.00010	.00015	.00021	.00028	.00037	.00049	.00063	.00082	.0011	.0016
10	.00013	.00019	.00025	.00034	.00045	.00058	.00075	.0010	.0014	.0021
20	.00017	.00023	.00031	.00041	.00053	.00069	.00090	.0012	.0018	.0031
30	.00021	.00028	.00037	.00049	.00063	.00082	.0011	.0016	.0026	
40	.00025	.00034	.00045	.00058	.00075	.0010	.0014	.0021	.0040	
50	.00031	.00041	.00053	.00069	.00090	.0012	.0018	.0031		
60	.00037	.00049	.00063	.00082	.0011	.0016	.0026			
70	.00045	.00058	.00075	.0010	.0014	.0021	.0040			
80	.00053	.00069	.00090	.0012	.0018	.0031				
90	.00063	.00082	.0011	.0016	.0026					
100	.00075	.0010	.0014	.0021	.0040					
110	.00090	.0012	.0018	.0031						
120	.0011	.0016	.0026							
130	.0014	.0021	.0040							
140	.0018	.0031								
150	.0026									
160	.0040									

$$S = \frac{\text{Applied Current}}{\text{Rated Current}}$$

Opto-Electronic Devices

The failure rate model for opto-electronic devices is as follows:

$$\lambda_p = \lambda_b * \Pi_T * \Pi_E * \Pi_Q \quad \text{Failures} / 10^6 \text{ Hours}$$

Table R.37 Environmental Factor

Environment	Π_E
G_B	1
G_F	2.4
G_M	7.8
S_F	1
M_P	7.7
A_{IC}	2.5
A_{UC}	3
A_{RW}	17
M_L	26

Table R.38 Quality Factor

Quality Level	Π_Q
MIL-Spec	0.5
Industrial	1.0

Table R.39 Junction Temperature

Device Type	T_J
Discrete LED	$T_A + 20 \ (^\circ C)$
LED Display	$T_A + 30 \ (^\circ C)$
Phototransistor	$T_A + 30 \ (^\circ C)$

T_A = Ambient Temperature

$$\Pi_T = 8.01 \times 10^{12} \ e^{-[8111/(T_J+273)]}$$

Table R.40 Base Failure Rate

Device Type	λ_b
Single LED	0.00065
Phototransistor	0.0015
Alpha-Numeric Displays 1 Character	0.00050

MICROELECTRONIC DEVICES

The general operating failure rate for microelectronic devices is as follows:

$$\lambda_p = \Pi_Q * (C_1 * \Pi_T * \Pi_V + C_2 * \Pi_E) * \Pi_L \text{ Failures} / 10^6 \text{ Hours}$$

Table R.41 Quality Factors

Quality Level	Π_Q
MIL-Spec Fully Compliant	2.0
Industrial	20.0

$C_1 = 0.01$: Circuit Complexity Factor for Devices Listed

Table R.42 Package Complexity Factors

Package Type	C_2
Hermetic DIPS with Solder or Weld Seals, Leadless Chip Carrier (LCC)	$2.8 * 10^{-4} (N_p)^{1.08}$
Hermetic DIPS with glass seals	$9.0 * 10^{-5} (N_p)^{1.51}$
Non-hermetic DIPS	$2.0 * 10^{-4} (N_p)^{1.23}$
Hermetic Flatpacks	$3.0 * 10^{-5} (N_p)^{1.82}$
Hermetic Cans	$3.0 * 10^{-5} (N_p)^{2.01}$

N_p is the number of functional pins ($3 <= N_p <= 180$)

Table R.43 Voltage Stress Factors

Technology	Π_V
CMOS, $V_{DD} < 12$ V	1.0
CMOS, $12 \le V_{DD} \le 20$ V	use Equ. (1)
CMOS/SOS & all Technology other than CMOS	1.0

Equ. (1) : $\Pi_V = 0.110(e^x)$

$$x = \frac{0.168 \cdot V_S (T_J + 273)}{298}$$

Table R.44   Device Learning Factors

Conditions	Π_L
(1) New device in initital production	10
(2) Major changes in design or process have occurred	10
(3) Extended interruption in production or a change in line personnel	10
(4) New and unproven technology	10
(5) If (1), (2), or (3) do not pertain	1

Table R.45   Environmental Factors and Case Temperature

Environment	Π_E	T_C
G_B	0.38	35
G_F	2.5	45
G_M	4.2	50
S_F	0.9	45
M_P	3.8	40
A_{IC}	2.5	60
A_{UC}	3.0	95
A_{RW}	8.5	60
M_L	13.0	60

$$T_C = \text{Case Temperature in } ^\circ C$$

Table R.46 Temperature Acceleration Factors For Integrated Circuits

Technology	Package Type	A
ASTTL, TTL	Hermetic Nonhermetic	4635. 5214.
LSTTL	Hermetic Nonhermetic	5794. 6373.
HCMOS	Hermetic Nonhermetic	6373. 9270.
Linear (Bipolar and MOS)	Hermetic Nonhermetic	7532. 10429.

Note 1. $\Pi_T = 0.1 \ (e^x)$

where

$$x = -A \left(\frac{1}{T_J + 273} - \frac{1}{298} \right)$$

A = value from above Table

T_J = device worst case junction temperature ($^{\circ}$C)

e = natural logarithm base, 2.718

T_J = $T_c + \theta_{JC} \ P$

T_c = case temperature

P = worst case power realized in a system application

Q_{JC} = 125 $^{\circ}$C/W For plastic (nonhermetic) IC's

    = 30 $^{\circ}$C/W For Hermetic IC's

Part III
Exercises

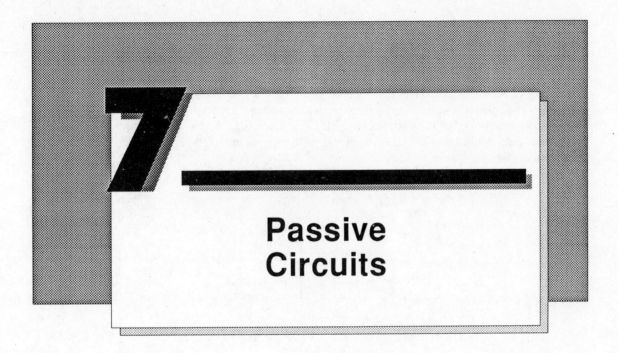

Passive Circuits

7.01

THREE-WAY SWITCHING CIRCUIT

Being able to switch a light on or off from two different locations, such as in a room having two entrances, is known as *three-way switching.* Single-pole double-throw (SPDT) switches are used in this application.

Draw the circuit diagram for the three-way switching arrangement. Following is the schematic diagram of a SPDT switch:

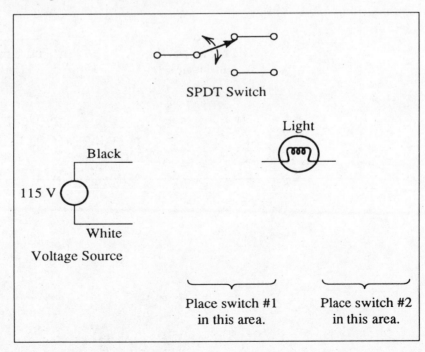

7.02

FOUR-WAY SWITCHING CIRCUIT

Three-way switching is fairly common; four-way switching is less so but is sometimes needed. *Four-way switching* enables a light to be switched on or off from three locations. In this arrangement, two of the switches are SPDT (see Exercise 7.01), while the third is a double-pole double-throw (DPDT) modified with internal connections, shown in the following figure.

Draw the circuit diagram for the four-way switching arrangement.

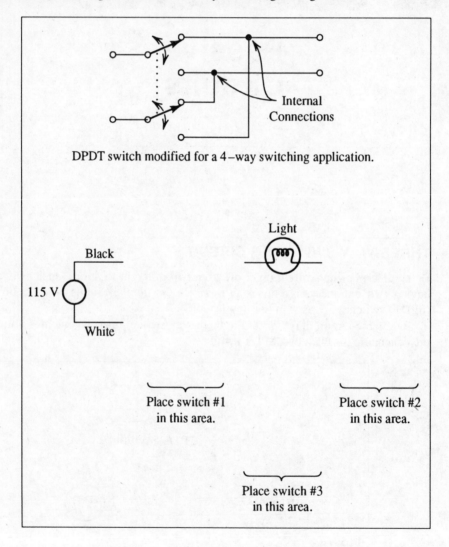

DPDT switch modified for a 4–way switching application.

7.03

ELECTRIC FOOT WARMER

Resistivity of Materials

Suggested by
R. W. Graff
LeTourneau College

You have taken a summer engineering job with a company that is developing a line of electric foot warmers. The heating element used in a foot warmer must be a coil of wire with a resistance of 2.7 Ω. You can choose from four kinds of wire: iron, copper, nichrome, and aluminum; the resistivities of each are listed below. The wire is available in 14, 20, 30, and 40 standard American Wire Gauge (AWG) sizes.

Material	Resistivity (Ω–cm)	Specific Gravity	Cost (cents/g)
Cu	$1.7 \bullet 10^{-6}$	8.94	0.65
Al	$2.8 \bullet 10^{-6}$	2.70	0.34
Fe	$97.8 \bullet 10^{-6}$	7.85	0.29
Ni Cr	$112.0 \bullet 10^{-6}$	8.25	1.54

A surcharge is incurred when using wire shorter than 2 m long because of the extra materials required to adequately dissipate the heat and the inherent added complexity of manufacturing the element.

Surcharge multiplier $= 1 + 1000e^{-10L}$,

where L is the length of the wire in meters. Thus the cost of making the element is the product of the materials cost and the surcharge.

Design the 2.7-Ω resistor that is the least expensive to produce.

7.04

TELEPHONE CABLE MAINTENANCE

Resistivity of Materials

Suggested by
Karl W. Berger
Lea + Elliott

A telephone subscriber has been without service since the last big rain storm. You, as the telephone company's chief engineer, suspect that a mud slide has crushed the underground cable running to the subscriber's residence, thus causing a short circuit. A precision resistance reading at the company's central office shows $351.2 \, \Omega$ between the line terminals. Drawing 249D573 shows a line 17.1 km long from the central office to the subscriber's phone. You know that all underground cables use AWG 20 copper conductors. With the telephone instrument on-hook, it represents an open circuit to dc measurements. Assume the only piece of equipment available is a precision ohmmeter, which has an accuracy of 0.1% or 0.1 Ω, whichever gives the greatest error.

Suggest a procedure for finding the exact location of the short using the smallest number of test holes. You may read the resistance at each hole. Each hole should be 2 ft in diameter.

7.05

WATER PURITY TESTER

Resistivity of Materials

Suggested by
R. W. Graff
LeTourneau University

A water treatment facility removes salt and other wastes from sea water that is used to cool a power plant and its machinery. To remove the debris, the facility passes the water through a series of filters and treatments until its level of purification reaches certain specifications.

Design a test station to test the water's conductivity to determine if the water has been processed correctly. Acceptable water quality occurs when conductivity falls below $1 \cdot 10^{-2}$ mho/m. At your disposal are a d'Arsonval meter movement rated at 50 mV, 1 mA; a 5-V dc power supply; and 5% resistors. The coil resistance of the meter has a tolerance of ±10%, and the needle deflection error is negligible. Provide the dimensions, the materials, and a drawing for constructing a conductivity measuring system for a 1-L test sample of water. Also provide a schematic of the circuit. The system should be sensitive enough so that conductivities of $1 \cdot 10^{-3}$ and $1 \cdot 10^{-1}$ will produce at least a 25% difference in meter deflection compared to $1 \cdot 10^{-2}$.

7.06

SHUNT DESIGN

Resistivity of Materials

Sometimes you need to measure a very large current without a meter capable of handling the current. You can do this using a *shunt,* that is, a very small high-current calibrated resistance that produces a measurable voltage drop when subjected to the test current.

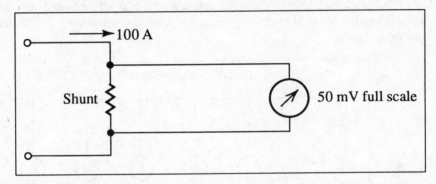

The shunt has four terminals, two large ones to circulate the heavy test current and two small ones across which the voltage is measured, and generally looks as follows:

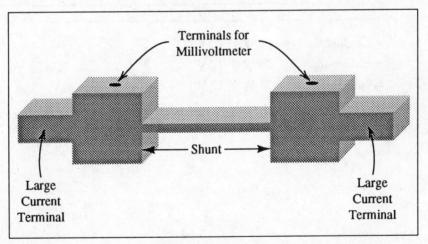

Make a drawing that shows all critical dimensions for a shunt to be made from copper (resistivity = $1.724 \cdot 10^{-6}$ Ω–cm) that will produce 50 mV between its voltage terminals when 100 A is flowing through it. If the machinist manufacturing the shunt makes a 1/2-mm error in the shunt's length, what percentage error is produced in the current measurement?

7.07

AUTOMOTIVE CIRCUIT PROTECTION

Current Divider/Parallel Equivalent Resistance

A fuse protects a circuit from catastrophic failure in the event of a short circuit. Fuses should be chosen to pass the largest expected normal operating current but to melt and open the circuit if a larger-than-normal current flows.

While restoring a 1957 Corvette, you discover that a fuse is missing. You have no information about its original value, so you trace the wiring and discover the following layout:

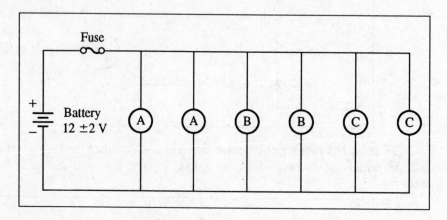

A = Headlight   60 W @ 12 V

B = Parking light   25 W @ 12 V

C = Taillight   25 W @ 12 V

Choose the proper fuse to protect the circuit. Use a standard fuse value from the approved parts list.

7.08

VOLTAGE DROPPING RESISTOR I

Voltage Divider Circuit

A friend has an old tractor manufactured in 1943. Although the electrical system was originally designed for a 6-V battery, your friend wants to convert the system to operate on a 12-V battery. However, he doesn't want to replace the horn, whose unique drone is useful for calling the dog. The old horn is designed to work on 6 V and it draws 5 A.

Specify a voltage dropping resistor to be connected between the horn and the 12-V battery, as shown in the following circuit:

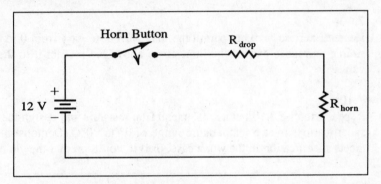

Note that the horn can be represented by a resistor.

Select the wire-wound resistor that most nearly allows 5 A to flow under the conditions shown and specify it with enough detail for your friend to be able to purchase it at the local electronics supply store. Five percent accuracy is sufficient.

7.09

VOLTAGE DROPPING RESISTOR II

Voltage Divider Circuit

Your friend has successfully converted his 1943 tractor from a 6-V to a 12-V electrical system (see Exercise 7.08), with one exception. He did not replace the lamp that illuminates the rear working area. When he connected the lamp to the old 6-V battery, he measured a current of 2 A in the circuit. He also observed that the lamp became too dim when the battery voltage dropped below 5.5 V. A notice on the lamp says that the absolute maximum voltage the lamp can sustain is 7.25 V. Your friend also made some measurements on the tractor's new electrical system and found that the battery voltage can vary from 12.0 to 14.0 V, depending on the engine speed. (When the battery is being charged, it may reach 14 V.)

Draw the circuit diagram that shows where to connect the series dropping resistor. Complete the design by specifying a resistor value that guarantees the voltage across the lamp will always be within the acceptable range of 5.5 to 7.25 volts. (This is a worst-case design.)

Assume the electronics distributor carries the line of resistors shown in Part 2 (data section). List all possible acceptable resistors in the order of increasing cost so your friend can minimize his cost in the event some resistors are not available.

7.10

REFERENCE VOLTAGE CIRCUIT I

Voltage Divider Circuits (no load and loaded)

Design a circuit that will provide a reference voltage at its output of 2.0 ±0.2 V. To develop the reference, you must use a simple two-resistor voltage divider. The voltage source is 5.0 V ±5%, and the only resistors available are standard values with 5% tolerance. The divider may not draw more than 1.0 mA from the source and the load draws less than 1.0 μA.

Choose the two resistor values and use Monte-Carlo analysis to show that the yield of acceptable dividers is at least 97% when the temperature is 25°C.

7.10a.
Repeat Exercise 7.10 but permit the load current to vary from 0 to 1.0 mA. Allow the source current to exceed the limit given above but restrict it to the smallest possible value.

7.10b.
Repeat Exercise 7.10 but use 1% metal film resistors and design the circuit to meet the specifications over a temperature range of 0° to 70°C. Demonstrate that your design meets specification in the worst case, or if it won't, predict the yield.

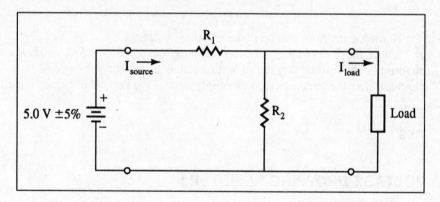

7.11

PRECISION VARIABLE VOLTAGE REFERENCE

Voltage Divider Circuit (unloaded)

A certain control used in a nuclear reactor requires very fine adjustments. A linear taper 10 kΩ carbon potentiometer is used to produce a reference voltage from 21 to 28 V for this control. A 48-V source is used to derive the reference.

Complete the design of the following circuit using nominal calculations and standard values of 5% tolerance resistors. Note that the reference voltage range must include the 21 V and 28 V extremes, but for maximum precision, the range should not exceed this amount by any more than necessary. When the potentiometer is at one end of its rotation, the reference voltage should be equal to or slightly less than 21 V; at the other end of the potentiometer rotation, the reference voltage should be equal to or slightly more than 28 V.

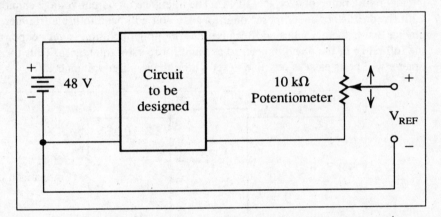

The design result should be (1) a circuit diagram that shows all component values and (2) a drawing of a dial face to be placed under the potentiometer knob showing the reference voltage produced (in 0.5-V increments) versus knob position. Assume the potentiometer has a mechanical rotation of 300° end-to-end.

7.12

VOLTAGE REGULATOR REFERENCE

Voltage Divider (loaded)

Xeno Motor Company is developing a new voltage regulator for the battery charging circuit. You are to design a voltage reference circuit. When the battery voltage is 14 V (battery voltage during charging), the output reference voltage is to be 8.8 ±0.05 V. The reference voltage is fed to a comparison circuit that has an input resistance of 1200 Ω. The reference circuit should draw as little current as possible.

Design the circuit using standard 5% resistor values. Use nominal resistor values at 25°C for the design.

7.12a.
Repeat 7.12 taking into account the tolerances of the resistors and their variation over a temperature range of –25° to +105°C. The tolerance on output voltage cannot be met with fixed resistors under these conditions, so you will need to use a trimmer potentiometer that is factory adjusted. Because the trimming operation is very expensive, use the full range of the potentiometer to account for resistor tolerances. Thus the potentiometer will be as easy as possible to set within the required tolerance.

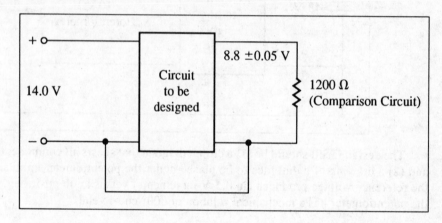

7.13

VOLTMETER DESIGN

D'Arsonval Voltmeter

Design a 30-V dc voltmeter that has an accuracy of better than ±2.5%. Use a d'Arsonval meter movement with a deflection accuracy of ±1%, a tolerance on the coil resistance of ±10%, and nominal ratings of 100 mV at 200 μA for full-scale deflection. Completely specify the series resistor to be used in the voltmeter circuit.

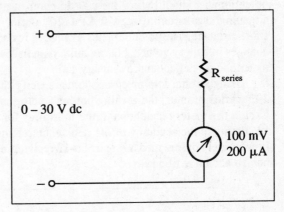

7.14

AMMETER DESIGN

D'Arsonval Ammeter

You are in charge of restoring antique electrical equipment at the Smithsonian Institution and are presently working on the control panel for Thomas Edison's 1887 power generating station that was at Niagara Falls. The restoration is going well, but the field current ammeter, which was designed to measure 0-50 A dc, has been destroyed; only its face remains. You have located a replacement meter that is nearly identical to the original except that its face is calibrated to measure 0-10 A dc. You determine that the resistance of the 0-10 A meter is 0.015 Ω. While you can easily insert the original face in the working meter, you will need to add a shunt resistor in parallel with the meter so it will read correctly. You decide to use a piece of solid copper wire across the terminals of the meter as the shunt.

Remembering that thin wires become very hot when carrying large currents, choose a size and length of solid copper wire to use for the shunt resistor that will provide the desired resistance and have a current density of no more than 1000 A/cm^2.

7.15

OHMMETER DESIGN

D'Arsonval Meter Circuits

Design an ohmmeter to measure resistance from 0 to 1000 Ω or more that will use a 50 mV, 1 mA, d'Arsonval movement with a needle deflection range of 120°. Use approximately 90% of the deflection range (120° to approximately 12°) to cover the resistance range of 0 to 1000 Ω. To power the meter, use a 1.5-V battery (1.2–1.6 V). The zeroing potentiometer should allow the user to compensate for variation in battery voltage by adjusting the meter to read 0 Ω (120° deflection) when the meter terminals are short-circuited. The potentiometer's range of resistance should compensate for possible changes in battery voltage but not more than necessary. When 1000 Ω is measured, the meter should deflect approximately 12°.

Using standard value components, specify the resistance and power rating of the 10% series resistor, the zeroing potentiometer, and the 1% metal film shunt resistor. Sketch the meter face showing the markings for 0, 1, 10, 100, 500, 1000, and 5000 Ω. Estimate the accuracy of the resistance reading if the user were to zero the meter immediately before measuring a 500-Ω resistor, assuming the d'Arsonval mechanism has an accuracy of ±1%.

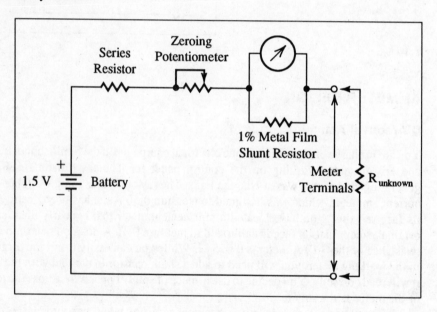

7.16

RESISTOR TOLERANCE TESTER

Wheatstone Bridge Circuit

Your company has received several shipments of resistors it suspects are out of tolerance. The resistors are supposed to be 220 Ω, $\pm 5\%$; you are to make random tests on them to verify those specifications. To do so, you need to design a testing device that will produce a "go/no-go" test result. The test circuit configuration is a Wheatstone Bridge circuit, as shown in the figure below.

The detector is a 50-mV galvanometer movement with a maximum allowed current of 50 μA at full-scale deflection in either direction. You decide that if you use 1% metal film resistors at nominal values for the bridge (resistors R_1, R_2, and R_3), the measurement should be sufficiently accurate. Choose R_1, R_2, and R_3 and mark the scale so the "OK" band is approximately one-fourth the full meter range. (Specify the meter current at the two demarcation points using nominal values of R_1, R_2, and R_3.)

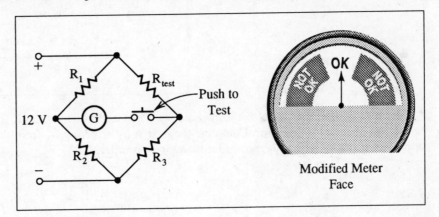

Modified Meter Face

7.17

REFERENCE VOLTAGE CIRCUIT

Voltage Regulation

Submitted by
R. W. Graff
LeTourneau College

Design a circuit to produce 3.5 ± 0.2 V at its terminals. This voltage should not change by more than 5% of its open-circuit voltage when a 100-Ω resistor is connected to its terminals. The circuit should be powered by as many 1.5-V flashlight batteries as needed. Assume the internal resistance of each battery is 2 Ω. Use nominal values of all components.

7.18

HANDWARMER DESIGN

Maximum Power Transfer

Suggested by
R. W. Graff
LeTourneau College

Design a handwarmer with a heating element consisting of a resistor to be held in your hand (or imbedded in your glove) and powered by a 6-V battery that you put in your coat pocket. Two wires will run from the battery in the pocket down the sleeve to the resistor. The circuit will be as follows:

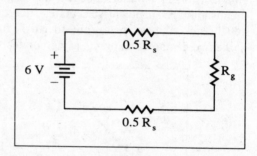

R_s is the total resistance of the two sleeve wires and equals 4.8 Ω; R_g is the resistance of the glove resistor. Complete the design by selecting R_g from the list of approved resistors so that the power into the glove is at least 1.87 W.

7.19

ELECTRICAL INVERTER FEEDER LINE

RMS Power Analysis

Submitted by
Karl W. Berger
Lea + Elliott

Your new motor boat is equipped with a low-voltage, dc power system. You would like to have a small inverter in the after-cabin to supply 120 V ac at 60 Hz to your VCR and wide-screen TV. These units use 500 W total and you've found a good inverter rated 700 W. The name plate reveals the following data:

700 W Intermittent
550 W Continuous
Nominal Output 120 V ac, 60 Hz
Input Range 22.7 - 28.6 V dc
Efficiency under load 82%
Fuse at 20 A

You measure 26 V dc under load at the battery terminals. The battery is located 10 m from the inverter. Specify the size of copper wire to feed the inverter.

7.20

HIGHWAY TUNNEL LIGHTING SYSTEM

AC Circuit Concepts

A contractor calls you for advice regarding installing lights in a highway tunnel. There are eight lights, each 400 W, to be fed from a single 220-V ±5%, 60-Hz source at one end. The lights are 60 m apart, and the distance from the source to the first light is 120 m. The last light won't work properly if the voltage across it is less than 200 V.

Design the wiring for the lights, including specifying the gauge of copper wire to be used and the proper fuse for the circuit. Use the same gauge wire throughout the circuit.

7.21

LOW VOLTAGE LIGHTING SYSTEM DESIGN

Submitted by
Parman E. Reynolds, P.E.
Reynolds & Sterling

A homeowner wants to install one string of 10 50-W landscape lamps evenly spaced at 10-ft intervals around his house. For safety reasons, he plans to use in the fixtures 24 V ac incandescent lamps that will be powered by a step-down, 120–24-V transformer. Circuits to the lights will be run in nonmetallic conduit underground. Assume the sizes of the conductors can vary between #12 and #8 AWG and the conductors operate at 25°C. Because incandescent lamps are sensitive to voltage drop, no more than a 1.0-V drop can be tolerated.

For one string of 10 50-W lamps evenly spaced at 10-ft intervals, determine the best wire size.

7.21a.
If the owner installs a string of five 25-W lamps fed from one end of the string and tapped in the middle to run to a 50-W fixture located 40 ft away, choose the wire size for the circuit. Assume the fixtures are located approximately 20 ft apart in the 25-W string and that the first fixture is 40 ft from the transformer.

7.22

HEAT SENSING SURVEILLANCE SYSTEM

RMS Power Analysis

As part of a sophisticated system of electronic surveillance for an art gallery, an electric current is circulated through a small wire embedded in a glass display case. The wire has a resistance of 100 Ω. The signal is produced by a special generator that produces the following waveform:

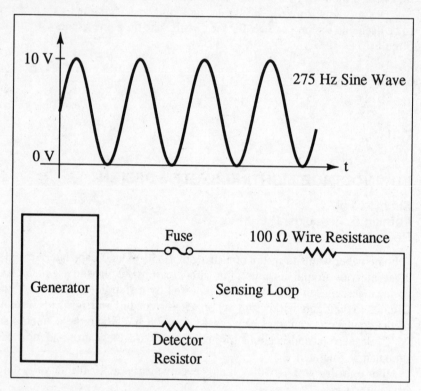

The receiver is simply a resistor that is heated by the passage of the current. A fuse is hidden in series with the sensor wire to complete the system so that if the resistance wire is bypassed, current flow will increase and blow the fuse. If the wire is cut or the fuse blows, current will stop flowing and the resistor will cool, thus setting off the alarm.

Complete the design of this system by specifying the appropriate fuse and detector resistor.

7.23

THEVENIN EQUIVALENT CIRCUIT

Suggested by
R. W. Graff
LeTourneau College

You are the junior engineer of the crew of a spaceship that has been heavily damaged in a meteor shower. To make emergency repairs, you need a dc source whose Thevenin equivalent voltage is 12 V and equivalent resistance is 10 Ω. You have found two spare sources, one 10 V with 5-Ω equivalent resistance and the other 4 V with 5-Ω equivalent resistance.

Using these two sources and some additional standard value 5% resistors, design the source you need. Nominal calculations are sufficient.

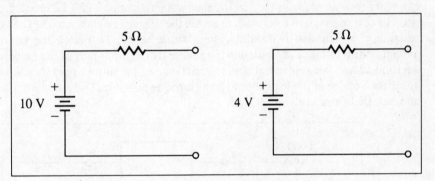

7.24

TWO-PORT MATCHING NETWORK

Thevenin Equivalents / Maximum Power Transfer

Submitted by
R. W. Graff
LeTourneau College

You are the only member of a group shipwrecked on a remote island who has any knowledge of electrical circuits. Therefore, it's up to you to rig a simple radio signal from items salvaged from the wreck.

You find a sonar unit from which you remove a signal source, which can be approximated as a 3-V high frequency source in series with a 1000-Ω resistor. You can make an antenna with the electrical characteristics of a 200-Ω resistor. Now you want to design a resistive network to match the source resistance to the load (antenna) resistance, which will make the system work better. To match the source to the antenna, the resistance presented to the source by the input port of the network should be 1000 Ω and the equivalent source resistance of the output port should be 200 Ω. Also, the voltage at the load should be as large as possible. Use standard 5% resistors and nominal values in your design.

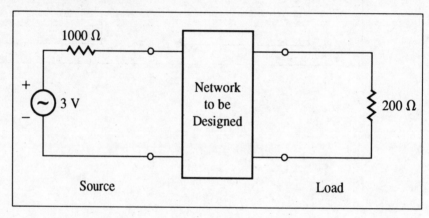

7.25

BOSE STEREO SPEAKER DESIGN

Maximum Power Transfer

Bose Corporation entered the audio speaker market in the late 1960s when its president became convinced he could design better speaker systems than he could buy. His first design was a five-sided enclosure incorporating nine 4-in. high-compliance speakers. The speakers each had an equivalent resistance of 8 Ω and were wired together in series-parallel. Because most amplifiers are designed to work best driving an 8-Ω load, Bose designed his system to match the output capabilities of such an amplifier. Each speaker produces sound power in direct proportion to the electrical power delivered to it, and the total sound power produced is the sum of the sound power produced by the individual speakers.

Design a circuit schematic showing how to wire the nine speakers to most nearly produce a load resistance of 8 Ω. You should equalize the power delivered to the individual speakers as much as possible.

7.25a.
Repeat your design showing the best wiring configuration of the nine speakers to most nearly produce an impedance seen by the amplifier of 4-Ω resistance. Don't let the difference in power dissipation between any two speakers exceed 50%.

7.26

STEREO SPEAKER DESIGN

Maximum Power Transfer/Equivalent Resistances

Submitted by
R. W. Graff
LeTourneau College

Your cousin has asked you to help him design a speaker system for his home. He has an audio power amplifier with which he wants to drive a stereo pair of 8-Ω speakers in every room, including the bathroom and basement. He has a total of seven pairs of speakers and needs to know how to hook up the seven 8-Ω speakers in series-parallel for each stereo channel so that the input resistance to the speaker network is between 7 and 16 Ω. In addition, each speaker should absorb nearly equal power. (*Note:* Do not use transformers.)

7.27

LOUDSPEAKER MATCHING

Impedance Matching/Maximum Power Transfer

Design an audio transformer circuit to connect three $8\text{-}\Omega$ speakers to an amplifier. Your circuit should have an input impedance of $8\ \Omega$ (to an audio signal in the frequency range 20 Hz to 15 kHz), and each speaker should absorb an equal amount of power. Specify the turns ratio.

7.28

BLACK BOX TEST PROCEDURE

Equivalent Resistance

Suggested by
Paul E. Gray
University of Wisconsin - Platteville

An electronics manufacturer has 10,000 identical black plastic cubes with four leads at the four corners of one side. The cube contains a network of 10% tolerance resistors arranged shown as follows, but there are no marks on the cubes to show which lead is #1:

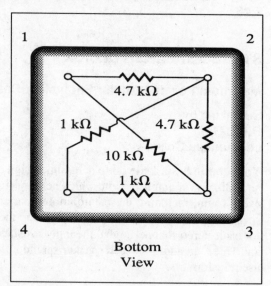

Devise a series of tests a technician can carry out on pairs of terminals with an ohmmeter to ascertain which terminal is #1. The terminal then can be marked with a white dot. The tests should be as simple as possible and should be stopped as soon as terminal #1 has been found for each cube. Be sure to specify the precision to which the measurements must be made.

7.29

ELECTROCARDIOGRAM RESISTOR NETWORK

Node Voltage Analysis

Suggested by
Paul E. Gray and Paul W. Garner
University of Wisconsin - Platteville

A living heart produces small electrical signals as it works. If the heart is diseased or injured, the signals will differ from those produced by a normal heart. A cardiologist often can determine what is wrong with a sick heart by carefully inspecting an electrocardiogram, which is a record of the signals appearing on a series of electrodes placed on the skin. To obtain an electrocardiogram, several electrodes are placed at specific locations. The signals they produce are added linearly by a resistor network, then amplified and recorded.

Design the resistor network and specify amplifier gain to produce a signal V as close as possible to

$$V = 1.3\ V_1 + 6.0\ V_2 + 2.7\ V_3 \quad \text{volts}$$

at the amplifier output. Assume the amplifier has an equivalent input resistance of 250 $k\Omega$. The individual resistors should be 5% tolerance and be between $10k\Omega$ and 100 $k\Omega$. All voltages are referred to the ground node.

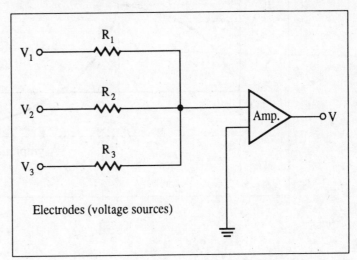

Electrodes (voltage sources)

7.30

WAVEFORM REALIZATION I

First-order Transients

Submitted by
R. W. Graff
LeTourneau University

Design a circuit to produce the voltage waveform $V_{out}(t)$, shown as follows, into a very high resistance load, which may be approximated by an open circuit. You may intervene to throw switches at specified times.

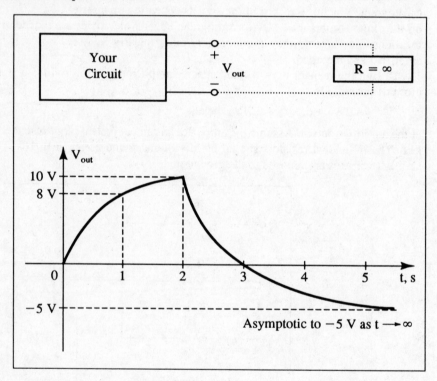

7.31

WAVEFORM REALIZATION II

First-order Transients

Submitted by
R. W. Graff
LeTourneau College

Design a circuit to produce the current waveform $i_o(t)$, as shown in the following figure, into a short circuit.

You may intervene to throw switches at specified times but don't use any ideal current sources.

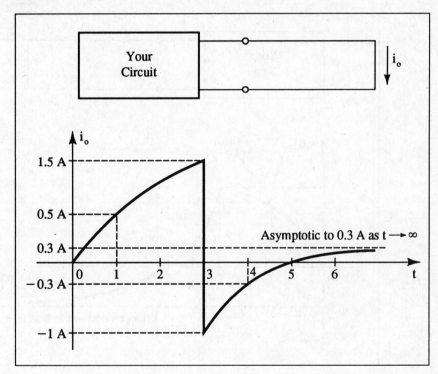

7.32

WAVEFORM REALIZATION III

First-order Transients

Submitted by
R. W. Graff
LeTourneau University

Design a circuit to produce the current waveform $i_o(t)$, shown in the following figure, into a short circuit.

You may intervene to throw switches at specified times but don't use any ideal current sources.

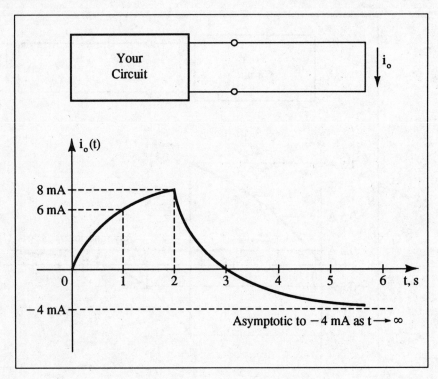

7.33

TEST PROCEDURE DESIGN

First-order Transients

Suggested by
Paul E. Gray
University of Wisconsin - Platteville

You are a consultant who is to develop a test procedure for the circuit module shown in the following schematic:

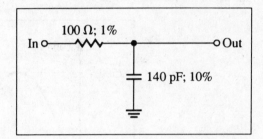

The procedure consists of applying a known signal to the input and observing the voltage waveform produced at the output. The manufacturer hopes that when the circuit is wired correctly and the components are within tolerance, the observed signal will fall within a certain range; otherwise, the signal will fall outside that range.

The network will be manufactured in large quantities. The test setup is to verify the accuracy of the 10% tolerance of the capacitor and check for correct wiring. Available equipment includes a single-channel oscilloscope with an input resistance of 1 MΩ and a capacitance of 10 pF and a square wave generator with an output resistance of 10 Ω.

1. Design a template for insertion over an 8 cm × 10 cm oscilloscope screen. The template pattern should provide test boundaries beyond which a produced module will be considered defective and therefore will be returned to the remanufacturing group.
2. Write instructions for the test procedure, assuming the operator generally knows how to operate the square wave generator and the oscilloscope. Be sure to indicate frequency and amplitude settings to be used.

7.34

TRANSIENT GENERATOR

Second-order Transients

Under some circumstances, the opening of a switch in an inductive circuit can cause a large voltage transient to occur. Because such transients can damage electronic components, protective devices should be installed on electronic equipment to prevent damage. For example, the Navy requires that all electronic equipment connected to shipboard 110-V ac lines be protected to withstand a 2500-V transient pulse on those lines, as shown in the following figure:

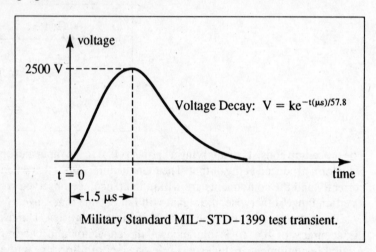

Military Standard MIL–STD–1399 test transient.

Design an R-L-C network with a push-button switch that will generate this second-order transient across a 1000-Ω resistive load when the button is pushed. If the transient generator is to be operated at a repetitive rate of 2/s or less, specify the minimum voltage, current, and power ratings of the components in the circuit, including the 1000-Ω load.

7.35

CAPACITOR TO REDUCE SPARKING

Second-order Transients

In the R-L circuit shown, the switch is closed long enough to establish a steady state and then is suddenly opened. To reduce the voltage that appears across the switch, it is proposed to connect a capacitor as shown in the following figure:

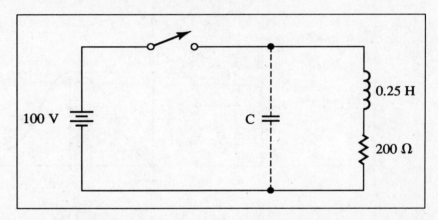

Choose a capacitor that will limit to 150 V the voltage appearing across the open switch contacts. What voltage rating should the capacitor have?

7.36

PULSE FORMING NETWORK

Second-order Transients

The following circuit is used as a pulse-forming network. Switch S is closed at $t = 0$. Sometime later, the voltage V_o reaches a maximum V_m.

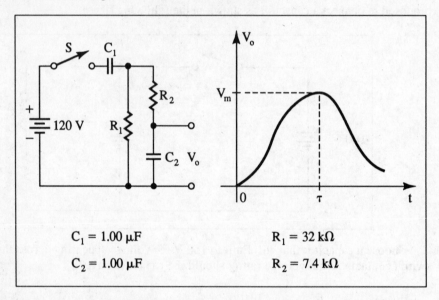

$$C_1 = 1.00 \ \mu F \qquad\qquad R_1 = 32 \ k\Omega$$
$$C_2 = 1.00 \ \mu F \qquad\qquad R_2 = 7.4 \ k\Omega$$

When the circuit was tested, V_m occurred at the proper time, but its value was too small.

Redesign the network to produce as large a peak as possible while still retaining the network's maximum at $t = \tau$. Don't alter the values of R_2 or C_1, but indicate values of R_1 and C_2 that will give the largest maximum.

7.37

CAMERA FLASH UNIT

Second-order Transients

Non-linear Circuit

You have been retained by a small business to design a super-bright camera flash unit that cave explorers can use underground. The basic circuit idea is shown as follows, as well as the flash-tube v-i characteristic:

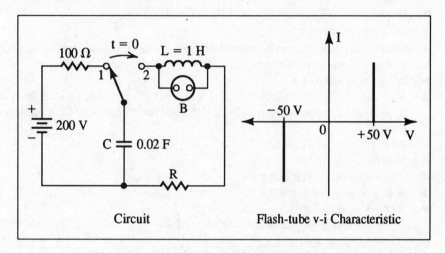

Circuit   Flash-tube v-i Characteristic

The switch ordinarily rests in position 1. A flash is initiated when the switch is thrown to position 2 and lasts until the current through the flash-tube B falls to zero.

Choose a resistance R so that the unit will produce a flash that lasts approximately 100 ms and which will not be repeated more than once every 10 s.

7.38

"HAUNTED HOUSE" SOUND PRODUCER

Second-order Transients

Submitted by
R. W. Graff
LeTourneau College

You are assisting a local service organization in setting up a "haunted house" for Halloween. You want to design a circuit that makes an eerie sound when a passerby steps on a switch. Accordingly, you determine that the sound can be produced from a voltage waveform that oscillates at 500 Hz and drops to 0.75 of its highest amplitude in $2 \text{ s} \pm 0.2\text{s}$ after the switch is stepped on. The amplitude of the oscillations should be at least 100 V peak for the first few milliseconds. Available for your use are a 12-V battery, a 50-mH inductor with negligible resistance, and a DPDT switch. Specify for purchase the other components needed in the design.

7.39

FAIL-SAFE BURGLAR ALARM

AC Circuit Concepts

Design a simple burglar alarm circuit with the following features:

- Monitors up to 30 windows or doors.
- Sounds an alarm if any window or door opens or if the wire to a sensor is cut (unless the system has been disabled with key).
- May be disabled with a key switch.
- Has an indicator light to show whether it is disabled.

You have at your disposal:

- Sensor switches that can be wired to be either open or closed when the door/window is closed. Switches have negligible resistance when closed.
- A 110-V ac siren, 3 A
- A siren relay (90 mA ±10% pull-in, 50 mA ±10% drop-out) that can be wired either normally open or normally closed. Coil resistance = 50 Ω. Contacts rated for 250 V, 10 A.
- Key operated DPST switch
- A 110-V, 1-W indicator lamp
- Any standard resistor you need
- A 110-V ac power source
- A 6.3-V ac transformer

Draw the complete circuit schematic, indicating clearly which components are housed in a central unit and which are distributed to the monitored locations. Show that your design will function at the limits of component tolerance.

7.40

PHASE SHIFT CIRCUIT

AC Circuit Analysis

The integral of a sine wave results in another sine wave but with a 90° phase lag. Many signal-processing applications require such an integration.

Design the following circuit to provide a 90 ±2° phase shift (lagging) into a 1000-Ω load. The combined load on the 500-Hz source must not draw more than 100 mA. Use approved 5% components.

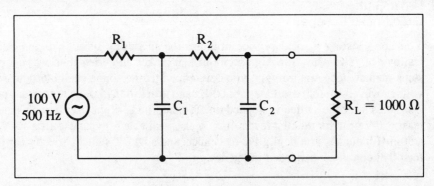

7.41

FLUORESCENT LIGHT BALLAST

Complex Impedance Voltage Divider

A fluorescent lamp is a relatively complex device, but during operation, it can be modeled as a simple resistor. For example, while lighted, a 20-W fluorescent lamp operates with 60 V across the tube. The voltage and current are in phase. To use this lamp on a 120-V ac, 60-Hz line, an inductor is placed in series as shown in the following figure:

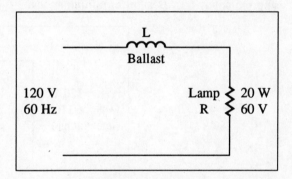

Design the circuit (specify the inductance value) so that the fluorescent lamp will operate at its rated power.

7.42

PHASE REFERENCE I

Phase Shifter

Your company is developing a line of single-phase power control devices. One such device requires a reference signal that lags the line voltage by 60°.

Design the following circuit to get as close as possible to the 60°-phase shift using a standard-value 5% resistor and a 10% standard capacitor. The input to your circuit is 120 V ac at 60 Hz. The output is to be a 60-Hz sine wave lagging the input by approximately 60°. Design for minimum cost.

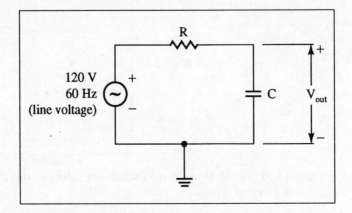

7.42a.
Repeat 7.42. The power control devices require a reference signal that lags the line voltage by 60 ±5°. This signal is to be derived from the line voltage, 120 V ac ±5%, at 60 Hz ±0.1%.

Design a circuit and demonstrate that your design will meet the requirements in the worst case.

7.43

QUADRATURE PHASE GENERATOR

An aircraft gyroscope requires two reference signals 90° out of phase (quadrature phase) and of equal magnitude. The aircraft power supply is 100 V ac at 400 Hz.

Design a circuit that will provide the appropriate signals. Note that both signals must be measured with respect to the aircraft common ground. To do this, you need to develop two circuits, one that will provide a +45° shift and the other with a −45° shift with respect to the input voltage. You may not use inductors. Your combined circuits should draw from 5 to 20 mA from the power source. Using nominal values, the phase angles must be within ±5%.

7.44

CAPACITANCE-METER DESIGN

Complex Impedance and AC Meter Circuits

Suggested by
Paul Leiffer
LeTourneau College

Design a meter to measure the capacitance of nonpolarized capacitors. You have at your disposal an ac meter movement that registers full-scale deflection for 1 mA rms and has 1000-Ω resistance and power at 120-V ac, 60 Hz. You want to measure capacitance in the range of 2 to 20 μF. Choose all components other than the meter from the approved parts list. Assuming the meter movement has a 120° displacement, draw the face of the meter and show the capacitance scale from 2 to 20 μF in 2-μF increments.

7.45

BELL TUNER'S HELPER

Resonance

Foundry workers are casting carillon bells. After the bells are cast, they tune the bells by either welding on small weights or grinding off part of the mass.

Design an electronic "ear" that can be used to determine frequency by the person who can't tell a B from a B flat. Refer to the following circuit and specify the passive components needed to have the meter response exhibit a sharp peak at 64.0, 76.1, 90.5, and 107.6 Hz. The peak should be sharp enough so that a deviation in frequency by 1.0% from resonance will cause the meter to fall to 90% of its peak reading. If suitable inductors are not in the approved parts list, fully specify ones that will work.

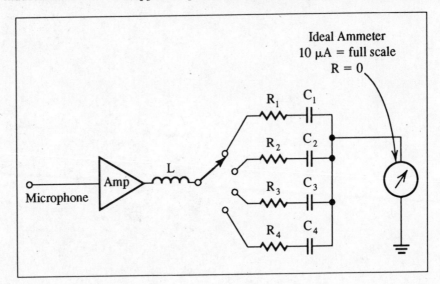

7.46

CONSTANT CURRENT CIRCUIT DESIGN

Resonance

Occasionally, you need to maintain a constant (or nearly constant) current through a component despite variations in the value of that component. The following circuit configuration supposedly allows a nearly constant current through R, regardless of its ohmic value, by taking advantage of the unique properties of a resonant circuit:

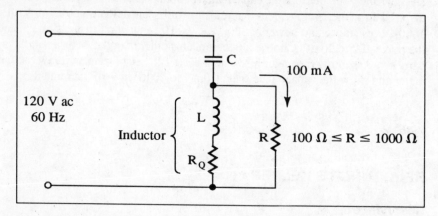

For the source shown, choose an L and C so that 100 mA flows through R for R $\approx 500\ \Omega$. Specify the values of L and C and their maximum voltage and current ratings. What is the minimum allowable $Q(\omega L/R_Q)$ for the inductor so that the load current will not deviate more than $\pm 5\%$ from 100 mA for values of R ranging from 100 to 1000 Ω?

7.47

BANDPASS FILTER

Resonance

As a customer service engineer for a manufacturer of heavy construction equipment, you have been sent to the desert to troubleshoot a drilling rig control system your company sold. The rig failed to work properly after it was set up in the field and the problem seems to be in a resonant circuit bandpass filter. The filter is designed to pass 1.0 MHz with a 3-dB bandwidth of 2 kHz. Unfortunately, the extreme daily temperature changes in the desert cause the center frequency of the resonant circuit to drift by about 500 Hz, thus causing the circuit to fail.

To reduce the filter's sensitivity to temperature, you decide to lower its Q and hence increase its bandwidth to 5 kHz. You elect to do this with a parallel resistor. Of course, the altered filter will have increased attenuation at 1.0 MHz, so you will need to add an amplifier to restore signal levels.

Complete this design change by specifying the parallel resistor R to be added and the gain A of the amplifier to be added.

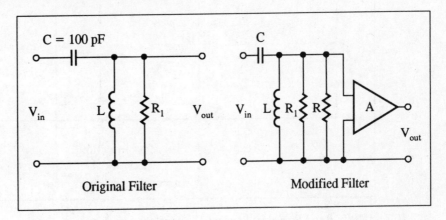

Original Filter   Modified Filter

7.48

LOSSLESS MATCHING NETWORK

Maximum AC Power Transfer

Submitted by
Charles Alajajian
West Virginia University

In communication circuits, the need often arises for a lossless matching circuit that does not dissipate energy but instead transforms impedance levels so that at terminals A, A′, the impedance is equal to the source impedance at a single input frequency. Following is a lossless matching network consisting of a series reactance arm and a shunt susceptance arm:

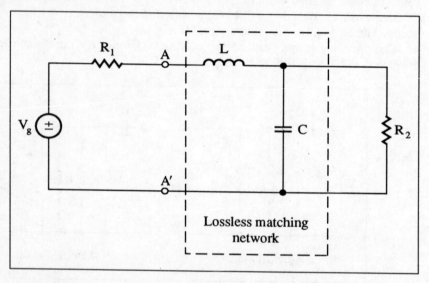

If $R_1 = 100\ \Omega$ and $R_2 = 150\ \Omega$, design a lossless matching network to achieve a match when the input sinusoid is set to a frequency of 2 MHz.

7.48a.

Design a second lossless matching network consisting of a single inductor and capacitor for $R_1 = 150\ \Omega$, $R_2 = 100\ \Omega$, and $f = 2$ MHz. (This network will require a different configuration of elements than that shown in the schematic.)

7.49

TONE CONTROL NETWORK

Complex Impedance and Frequency Response

Your company is designing an audio system for sale in a specialized market. You are to specify all component values for the following tone control circuit:

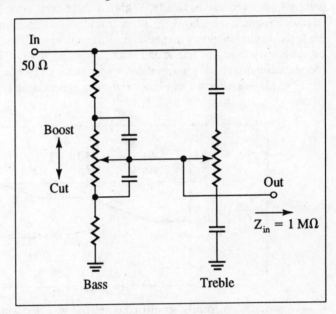

This circuit lies between a tape player and the amplifier section. The tape player output is equalized to compensate for the frequency response of the record-playback characteristic and has an output impedance of 50 Ω. The amplifier has an input impedance of 1 MΩ. The bass and treble controls in the circuit must be able to cut or boost by as much as 20 dB and must be continuously adjustable. The treble is to control frequencies above 2 kHz, the bass frequencies below 200 Hz. The bass and treble circuits can be analyzed separately. The circuit should have a gain of about −20 dB for midband frequencies.

7.50

RIAA EQUALIZER

Complex Impedance and Frequency Response

When Thomas Edison invented the phonograph, the process of converting sound energy to produce oscillations of the recording stylus and the complementary playback process were all achieved mechanically. Within a short time after Lee deForest invented his "audion" vacuum tube, however, other inventors succeeded in designing audio amplifiers to use in conjunction with phonograph recordings. Although playback volume could be made very loud, the music and voices sounded "tinny." Experiments revealed that the recording/playback process did not reproduce all frequencies with the same intensity. The playback/record amplitude ratio was approximately as shown in the following figure:

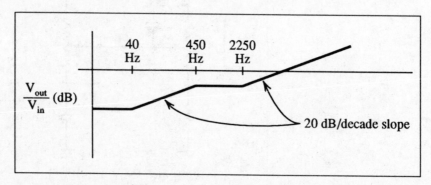

To produce uniformly amplified output, it was necessary to produce a filter that could compensate for the curve shown so that the overall response was "flat" to well above 2250 Hz. This compensating filter was called RIAA (Record Industry Association of America) equalization.

Using passive elements only, design an RIAA equalization circuit. Assume your circuit will be followed by an appropriate amplification circuit with an input resistance of 100 kΩ. (In a typical phonograph amplifier, this equalization would be integrated within the amplification stages with the appropriate gain.) Design for nominal values.

7.50a.

Your company wants to manufacture a passive RIAA equalization circuit for use with modern amplifiers that were not designed to be used with old phonograph records. The input impedance must be greater than 2000 Ω at all audio frequencies (20 to 20,000 Hz) and work with a 1000-Ω load. The pole and zero frequencies must be within 10% in the worst case, using standard components and over a temperature range of 0° to 70°C. The best design will be the least expensive one.

7.51

CROSSOVER NETWORK

Bode Plots

When an audio amplifier is used to drive two speakers—a woofer and a tweeter (8-Ω resistive impedance each)—a crossover network is required to channel the low frequencies to the woofer and the high frequencies to the tweeter.

Design the following crossover network to provide a crossover at 2 kHz:

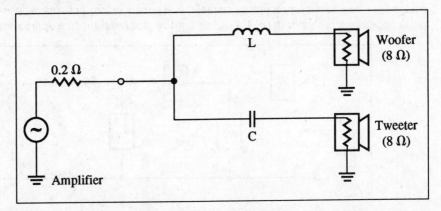

The proof of your design should be Bode plots of the power delivered to the speakers. These plots should include the power delivered to each speaker as well as the sum of the power delivered to both. The best design will be the one that most nearly allows constant total power to be delivered over the entire frequency range. Assume the power output of the amplifier is 5W. If the components in the approved parts list do not have appropriate voltage and current ratings, specify the required ratings.

7.52

DRIVING POINT IMPEDANCE

Impedance Networks

Submitted by
Paul E. Gray
University of Wisconsin - Platteville

Your employer builds commercial audio equipment for recording studios. As part of a design for a new low-cost recording console, you need a ladder network whose driving point impedance is shown in the following log-log plot (straight-line approximation):

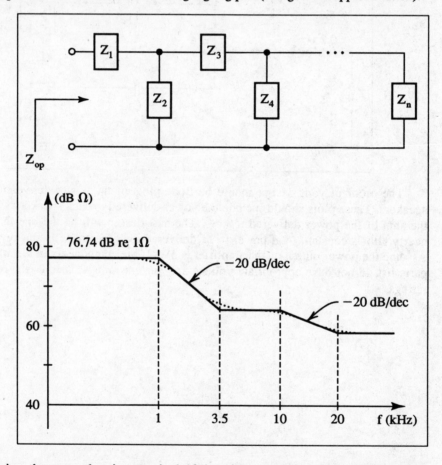

Design the network using nominal 10% resistors and capacitors. To compare results, plot the impedance of your network on the same graph as the desired impedance.

7.53

CABLE SIGNAL SPLITTER

Impedance Matching

A particularly useful small signal device is a "signal splitter" that can be inserted into a cable TV system, for instance, so that two TVs can receive input from a single cable. The cable system is designed to feed its signal to a single TV, which has a 75-Ω input impedance. If two TVs are simply hooked up in parallel, their combined impedance will be 37.5 Ω, which is not the value that the cable is designed to work with.

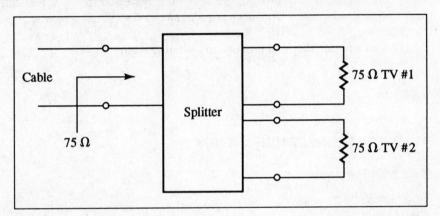

Design a splitter to be inserted into the cable that will allow two TVs to operate in parallel while still presenting 75-Ω impedance to the cable system.

7.54

IMPEDANCE NETWORK FOR A HOVERING SKATEBOARD

Submitted by
Paul Leiffer
LeTourneau University

Our design for a hovering skateboard requires a passive two-terminal network whose impedance is as large as possible at 120 rad/sec and is equal to 20 Ω at dc. Develop such a network using approved components.

7.55

STABLE SPEED CONTROL SYSTEM

Complex Admittance

Submitted by
R. W. Graff
LeTourneau College

You are working on a design team that is developing a speed control system for a motorized manufacturing process. To make the system responsive and stable, you must design an admittance $Y(s)$ that has a zero at $s = -6 \cdot 10^3$ and a pole at $s = -2 \cdot 10^3$. Furthermore, $Y(\infty) = 2 \cdot 10^{-3}$ Siemens.

Design a circuit that will produce this admittance function.

7.56

TWO-PORT IMPEDANCE MATRIX

Two-port Parameters

Suggested by
Paul Leiffer
LeTourneau College

Design a two-port network that will exhibit the following impedance matrix:

$$Z = \begin{Bmatrix} 300 & -700 \\ 400 & * \end{Bmatrix} \qquad * - \text{your choice (any value)}$$

Using passive components, come as close as you can using nominal values. Units are in ohms.

7.57

VOLTAGE TRANSFER FUNCTION

Laplace Transforms and Transfer Functions

Suggested by
Paul Leiffer
LeTourneau College

Design a circuit that has the open-circuit voltage transfer function characteristics shown in the pole-zero diagram below:

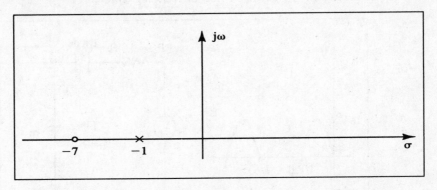

The magnitude of the transfer function should be 0.1 at the origin.

7.58

POWER SUPPLY RIPPLE FILTER

Fourier Series Analysis

Almost every electronic instrument incorporates a power supply, that is, a circuit that converts the power supplied from a wall outlet into a form that can be used by the instrument. A typical power supply contains a rectifier to convert alternating current to direct current and a filter or regulator to make the dc voltage nearly constant. The small variations in a nearly constant dc voltage are called *ripple*, which is undesirable.

The following circuit shows a filter for reducing ripple:

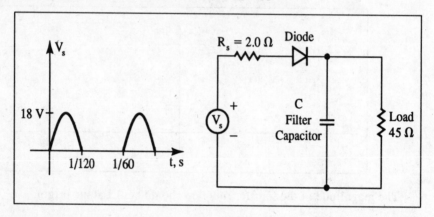

The rectifier produces a half-wave rectified sinusoidal voltage as shown but has 2.0-Ω source impedance. The instrument itself, which uses the dc, can be represented as a 45-Ω load. Assume an ideal diode that conducts with no loss in the forward direction and is an open circuit when reversed. Choose the least expensive capacitor that will guarantee to reduce ripple voltage to less than 0.2 V peak-to-peak.

7.59

BANDPASS FILTER DESIGN

Fourier Series Analysis

Submitted by
R. W. Graff
LeTourneau College

Design a passive bandpass filter that will extract the fifth harmonic from the following waveform:

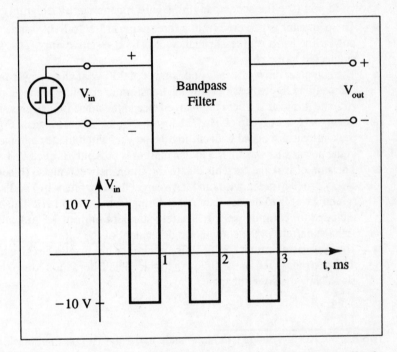

The resultant output should be approximately a sine wave, that is, the amplitude of the 3-kHz and 7-kHz components in the output waveform should be at least 10 dB less than the amplitude of the 5-kHz component.

7.60

PASSIVE-FILTER DESIGN

Submitted by
Joseph H. Wujek
University of California at Berkeley

As an electrical engineer in the EMC (Electromagnetic Compatibility) Engineering Department, you have been assigned to solve the following problem.

The company's new personal computer needs a power-line filter in order for it to comply with FCC Rules, Part 15J, for Class B devices. The following information is known to you (see the accompanying figure for the circuit diagram):

- The computer is powered from a three-wire 115-V, 60-Hz source. Assume the power line has a source impedance of 0.1 Ω, resistive only. The green wire is connected to the chassis.

- The computer draws a steady-state power of 230 W at unity power factor. The accompanying figure doesn't show the transformer coupling to the dc power supply. Assume this load is a very high impedance at the noise frequencies shown in the following table relative to 60 Hz. The inrush current is not to exceed 6 A. (*Inrush* is the peak current associated with closing the power switch to the computer at any point in the power sine wave.) The inrush current is presently limited to 3 A.

- The sum of leakage currents due to the filter, between phase (black wire) and safety ground (green wire) and between neutral (white wire) and safety ground (green wire) is to be less than 2.0 mA rms from dc to 5 kHz. This requirement is included for compliance with a safety standard limit of 3.5 mA and allows for leakage current in the computer independent of the filter.

- The equivalent circuit for the noise generator has a source resistance of 1 kΩ and the spectrum given in the following table. The filter is to reduce these noise levels as shown also in the table.

Table 7.60. Noise Voltage Characteristics and Filter Requirements.

e_N Noise Voltage (rms)	Frequency Band (MHz)	e_{NF} Filtered Noise Voltage (rms)
3.5 mV	36 ± 2	< 125 μV
820 μV	0.45 to 60	< 125 μV

1. Express the noise voltages of the table in decibels referenced to 1 microvolt.
2. To satisfy the requirements, what is the minimum value of attenuation in dB needed in each frequency band of the table?

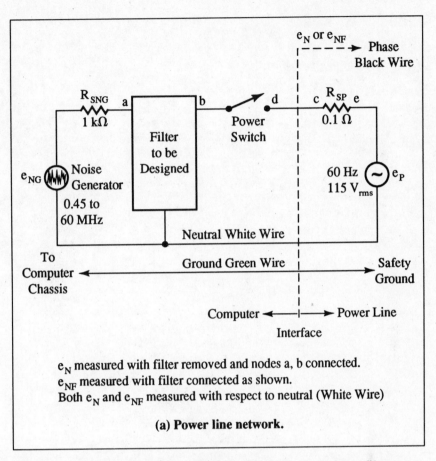

e_N measured with filter removed and nodes a, b connected.
e_{NF} measured with filter connected as shown.
Both e_N and e_{NF} measured with respect to neutral (White Wire)

(a) Power line network.

3. Examine the requirements and determine if it is possible to design a filter to meet them. If it is possible, design the filter; if it's not, show by analytical argument why not.
4. If you designed a filter in (3), show by analysis (with or without the LISN of figure 2) that it meets the requirements. Note that this includes surge and leakage as well as attenuation.

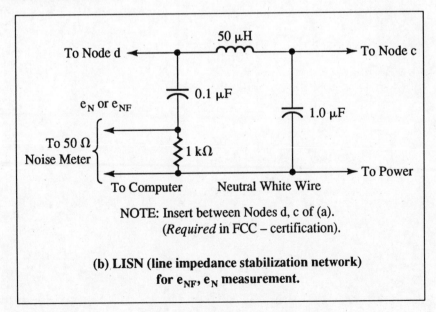

NOTE: Insert between Nodes d, c of (a).
(*Required* in FCC – certification).

**(b) LISN (line impedance stabilization network)
for e_{NF}, e_N measurement.**

8

Electronics

8.01

AC AMMETER

Rectifier Circuits

You are involved in a project to provide electrical power for a remote hospital by damming a local stream and using a water turbine. For the hospital equipment to operate correctly, the electrical generator powered by the turbine must be adjusted to deliver 30 A ac. Unfortunately the 0–50-A ac ammeter intended to register current was damaged beyond repair in shipping. However, you are able to salvage a dc meter from a scrap vehicle, and by making a few measurements, you determine that the meter reads full scale when 10 mA dc passes through it, at which time the voltage across the meter terminals is 100 mV dc. The meter reads average current.

Design a circuit that allows you to use the meter movement to read 0–50 A ac.

8.02

ZENER DIODE REGULATOR DESIGN

Submitted by
Hobart F. McWilliams
Montana State University

Design a Zener diode regulator to meet the specifications shown on the following circuit. Assume the diode, to regulate properly, must have a current I_Z of at least 1 mA.

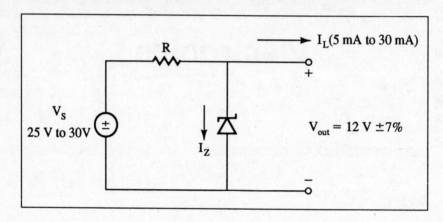

8.03

DISCRETE COMPONENT POWER SUPPLY

You need a dc power supply to run electronic equipment at a remote location. The load requirements are +12 and −12 ±0.6 V, at 1 A. Because the only power available is a portable generator, your source will be 120 V rms ±10%, 60 ±1 Hz. You have a 25.2-V, center-tapped transformer. You may use only discrete devices from the approved parts list.

8.03a.
Design the power supply so you can guarantee that at least 90% of the power supplies assembled will operate within specifications over the temperature range of −55° to +100°C.

8.03b.
Design the power supply so that the reliability (components only) is at least 0.99 for 1 yr of operation.

8.04

POWER SUPPLY CURRENT LIMITER

Design a current-limiting circuit to be added to the power supply design from Exercise 8.03. Again, you may use only discrete components. The current must be limited to no more than 1.2 A.

8.04a.
Design the current-limiting circuit so that the reliability of the entire power supply design meets the requirements of exercise 8.03b.

8.05

VARIABLE REGULATED POWER SUPPLY

Submitted by
Paul Gray
University of Wisconsin - Platteville

Design a variable-voltage regulator that can replace several different fixed-voltage regulators now being produced. The proposed schematic, shown as follows, uses an LM317T three-terminal adjustable voltage regulator IC. (You may make minor changes in the schematic.)

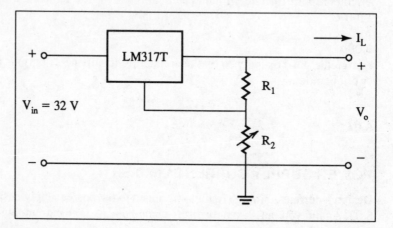

The fixed resistor is a standard 5% resistor. The input dc supply is 32 V at no load and develops a 2-V ripple at a 1-A load. Because of heat limitations, the regulator IC must not dissipate more than 15 W.

Design the circuit and write up a specification sheet for your design showing the allowable operating region.

8.06

COMPENSATED LED PILOT LIGHT

Large Signal DC Amplifier

You are designing an ambient light compensation circuit for an LED pilot light for a piece of field gear. The LED must be bright enough to be seen in open sunlight; however, as night falls the intensity should decrease correspondingly so that the LED just barely glows in the dark (to prevent its interfering with the operator's night vision).

In sunlight (greater than 5 mW/cm^2 of incident radiation), approximately 50 mA should pass through the LED. At night (0.05 mW/cm^2 incident radiation), approximately 5 mA should pass through the LED. The LED has approximately 2.1-V drop across it in this current range.

Your employer thinks an adequate compensation system can be produced using a photo-d'Arlington transistor, one or two fixed resistors, and the 15-V power supply of the field gear. Design such a circuit using approved components at nominal values.

8.07

ENERGY EFFICIENT PILOT LIGHT

Suggested by
Ladimer S. Nagurney
University of Hartford

Your employer manufactures energy-efficient electronic equipment. You are to design an LED pilot light circuit. This circuit must work directly off 110 V ac (no transformer) to light an LED and can use no components that waste power (no resistors or inductors); however, it must provide an rms current through the LED of 80% of the maximum rating of the device. Use the visible LED from the approved parts list.

8.08

WORST-CASE JFET BIAS DESIGN

Submitted by
Hobart F. McWilliams
Montana State University

You are to develop a dc bias circuit for a common source JFET amplifier that will restrict the variation in drain current to a specified maximum deviation between worst-case limits of the device parameters I_{DSS} and V_P.

A good bias design will be one that ensures I_D and V_{DS} will always be within a certain range of their nominal values, independent of the values of I_{DSS} and V_P. These parameters vary with temperature; more importantly, however, they vary considerably between different units belonging to the same device type. If the bias design is not satisfactory, then parameter variations between different units belonging to the same JFET type will severely degrade reliability of operation and manufacturing acceptance.

Design the bias circuit for an n-channel JFET (2N5485) that will restrict maximum quiescent drain current variation over worst-case limits to 1 mA. As part of your design, research manufacturer data sheets to obtain the necessary data to determine the worst-case bias points for the circuit.

8.09

COMMON-EMITTER AMPLIFIER I

You are repairing a piece of electronic equipment. Although you have partial knowledge of the circuit, the circuit diagram is missing. The amplifier circuit is so badly damaged that you can't see the component values to replace them. Thus, you will have to redesign the circuit.

The gain of the circuit is to be 14 dB and is a capacitor-coupled, common-emitter amplifier using a 2N3904 transistor. It was biased with a collector current of 10 mA and with collector-emitter voltage of 6 V. The supply voltage is +12 V. The input impedance must be greater than 1 kΩ and the output impedance must be less than 50 Ω. Use a 2N3904 transistor and resistors from the approved parts list.

8.10

COMMON-EMITTER AMPLIFIER II

Submitted by
Paul Leiffer
LeTourneau University

You are designing a radio receiver made entirely of discrete components. For the audio output stage, you need to design a common-emitter amplifier with a Q point of $I_C = 6$ mA and $V_{CE} = 8$ V for good linear amplification. The amplifier should have a voltage gain of at least 20 over the frequency range of 20 Hz to 20 kHz. You have a +15-V dc supply available. The input signal comes from a source with an output impedance of 200 Ω. The load on the amplifier is 50 Ω. Both input and output are to be capacitively coupled.

8.11

TRANSISTOR AMPLIFIER

As an electronics engineer for a toy company that is going to introduce a new talking doll, you are to design a single-stage transistor amplifier with a voltage gain of 20 ±3 dB into an ac-coupled load of at least 750 Ω. The signal source has a source impedance of 500 to 750 Ω and produces audio signals as large as 0.10 V peak-to-peak. You must guarantee a 3-dB frequency range of at least 300 to 4000 Hz. There can be no clipping of the signal. The power source is a 3-V battery pack, thus you must minimize power drain. You may use transistors, resistors, and capacitors from the approved parts list.

The best design will minimize costs, failure rate, and quiescent current, while meeting performance requirements. Cost and failure-rate calculations should include components only and must include the coupling capacitors. The temperature range is 0° to 70°C. The quiescent current is to be calculated for the entire circuit at worst case.

8.12

PUSH-PULL AMPLIFIER

An emitter follower amplifier you designed for the talking doll in Exercise 8.11 works satisfactorily. However, it draws too much power from the batteries because the bias current in the emitter follower is on continuously. You recall a discussion of push-pull amplifiers in a college electronics class and decide to design a new driver that will reduce the quiescent current to zero.

The doll has two 3-V battery packs. The toy company has determined that the best design would have minimum product of cost times failure rate. You can use only approved discrete components to reduce the effects of crossover distortion.

8.13

SEISMOMETER

DC Voltage Amplifier

Your neighbor, a geologist, has asked you to design a portable seismometer to help with seismic exploration. The detector is a moving coil within a permanent magnetic field. The signals the geologist wants to detect produce 300-μV sine waves at 5 Hz. You have a 500-μA, 50-mV dc meter you can use for the readout device. The system will be powered by a pair of 1.5-V D cells, resulting in a 3-V power supply for your circuit.

Use discrete components from the approved parts list to design a circuit that will provide a quiescent current of 250 μA in the meter with no signal and will produce a 100-μA sine wave deflection when the signal is present. The detector coil resistance is 200 Ω and you may have a dc current of as much as 10 μA without disturbing its accuracy. A design using nominal and typical values is sufficient.

8.14

IMPEDANCE MATCHING AMPLIFIER I

Suggested by
Paul Leiffer
LeTourneau University

Design a one-transistor circuit to match a high-impedance source to a low-impedance load. The circuit should have a voltage gain greater than 0.9 and an input impedance as large as possible (at least 1 kΩ) with a 16-Ω load. The output impedance must be less than 16 Ω. Assume a +12-V power supply is available.

8.14a.
Repeat 8.14, except that the input impedance must be greater than 20 kΩ. You may need more than one transistor.

8.15

IMPEDANCE MATCHING AMPLIFIER II

Suggested by
Paul Leiffer
LeTourneau University

Your client requires an unusual one-transistor amplifier, one that ideally will have an input impedance of less than 100 Ω, an output impedance as large as possible, and a no-load voltage gain of at least 50.

Determine the input impedance, output impedance, open-circuit voltage gain, and short-circuit current gain so that the client can choose the design that best meets its needs. For impedance calculations, assume the source resistance is 25 Ω and the load resistance is 1 kΩ. Use nominal and typical values of component parameters.

8.16

POWER AMPLIFIER DESIGN

Submitted by
Paul Leiffer
LeTourneau University

Design a transformerless audio power amplifier with at least 38% efficiency. A ±30-V dc supply is available at the client's site. The amplifier should be capable of supplying at least 5 W to a 16-Ω load at 1 kHz. The input impedance should be greater than 100 Ω and the voltage gain should be greater than 0.8. Provide a circuit schematic, a parts list, and the cost of parts.

8.17

FREQUENCY RESPONSE

Submitted by
Paul Leiffer
LeTourneau University

Design a capacitor-coupled common emitter amplifier with fixed bias and an emitter resistor for stability. Values used for C_C, C_E, and C_S should yield critical frequencies (poles) at 50 Hz, 1 kHz, and 50 kHz.

8.18

10x AMPLIFIER

Submitted by
R.W. Graff
LeTourneau University

Design an amplifier circuit using a 741 operational amplifier that will produce a 1-kHz sinusoidal signal of 10 mV ±10% peak-to-peak when it has an input signal of 1 mV peak-to-peak applied to it from a signal generator with 600-Ω source resistance. Assume a ±12-V power supply is available.

8.19

ANALOG COMPUTER DESIGN

Using 741 op-amps and a ±15-V power supply, design an analog computer circuit to solve the following simultaneous equations:

$$x = 6\,V_1 + 2\,y$$

$$y = -8\,V_2 + 10\,x$$

Show clearly where inputs V_1 and V_2 are to be applied and where solutions x and y may be monitored.

8.20

EMG SIGNAL AMPLIFIER

High-Gain Differential Amplifier

Suggested by
Paul Leiffer
LeTourneau University

Design a differential amplifier for EMG (electromyogram) signals recorded from limb muscles. The pick-up electrodes exhibit about 5-MΩ source impedance. These signals are on the order of 10 μV with frequencies from 10 Hz to 2 kHz. Design for a voltage gain of 100 dB at midband using a ±12-V power supply.

Amplifier Specifications

Gain:	100 dB, midband
Frequency response:	10 Hz to 2kHz (3 dB frequencies) 20 dB/decade roll-off outside band
Input impedance:	10 MΩ
Output impedance:	Less than 50 Ω

8.21

"NEGATIVE" RESISTOR

Submitted by
R.W. Graff
LeTourneau University

Using a 741 operational amplifier and other components, design an active circuit that exhibits a negative input resistance of -1000 ± 100 Ω. Assume a ±12-V regulated power supply is available. Only small signals will be applied to the circuit. Indicate the frequency range over which the circuit can be expected to function within the specifications given.

8.22

ACTIVE FILTER DESIGN I

Submitted by
Paul Leiffer
LeTourneau University

Your client is developing a remote audio speaker system that uses house wiring to send signals to speakers at distant locations. To implement the system, the client needs an active filter with the following voltage gain versus frequency response:

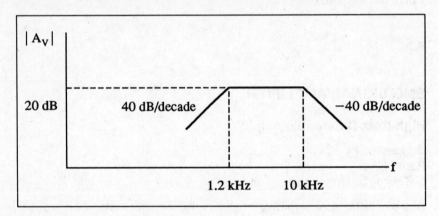

Design a suitable filter, specifying all components and using nominal values.

8.23

COMPUTER-AIDED FILTER DESIGN

Suggested by
Paul Leiffer
LeTourneau University

You are exploring the possibility of manufacturing low-cost PC software for electrical engineers to aid them in their design of active filters. The software would accompany the latest edition of *The Filter Designer's Handbook.*

Prepare a computer program to aid in the design of low-pass, high-pass, or band-pass maximally flat filters based on the following circuit:

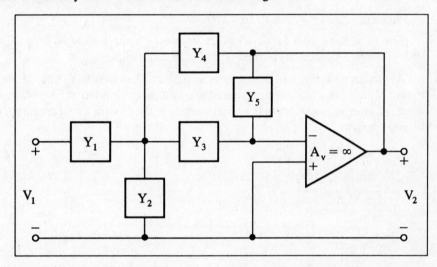

The program should accept as inputs the filter type and the desired 3-dB frequencies. Program outputs should include the number of stages necessary and the type and value of each of the five passive components for each stage, up to four stages. Demonstrate your program using at least two filter designs.

8.24

ACTIVE FILTER CONTEST

Submitted by
Michael E. Parten
Texas Tech University

As part of an IEEE contest to choose an outstanding student circuit designer, you are to design, analyze, and simulate an active filter system with the following specifications:

Input:	$1\ k\Omega$; 100 mV rms
Output:	$50\ \Omega$ load; 10 V rms
Frequency range:	dc to 10.0 kHz

Gain to be flat within 1 dB over given frequency range with rapid drop off (no less than 60 dB/decade) outside of the frequency range.

Assume a ±12-volt power supply is available. The circuit analysis of each filter section of the system must be compared to a computer simulation, including loading effects of other sections. Include a cost estimate of the system based on the parts costs for your design.

8.25

PHYSIOLOGICAL AMPLIFIER

Feedback Amplifier

As part of your job designing electrophysiological instrumentation for a large medical equipment manufacturer, you are to design a device to record electrical activity from single cells in bees' eyes. You soon discover that because the recording electrodes are so small, they have exceedingly high source impedance. You realize that your instrumentation must exhibit an input impedance of greater than 10^9-Ω resistance in parallel with no more than 0.5 pF capacitance. The input impedance of the instrument you have, however, is $10^7\ \Omega$ in parallel with 10 pF. A co-worker tells you she has seen a feedback circuit that will do the trick and sketches it on a napkin, as follows:

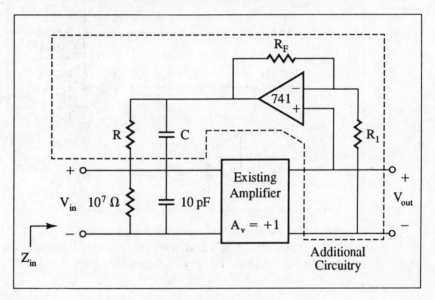

Complete the design by specifying component values and tolerances to achieve the desired impedance without causing circuit instability.

8.26

RMS VALUE OF SINUSOID

Submitted by
Dennis G. Smith
The University of Alabama at Birmingham

The following circuit is to provide an output dc voltage that is equal to the rms voltage of an input sinusoid:

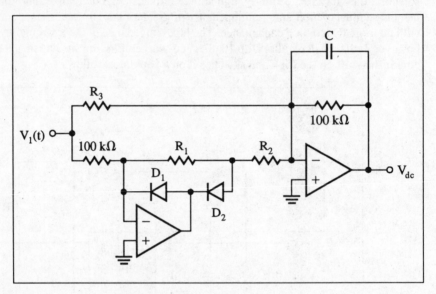

Complete the design for sinusoidal input voltages with a peak value of less than 15 V. Assume a ±18-V power supply is available, the operational amplifiers are ideal, and C is large enough to reduce the ripple to a negligible value.

8.27

AUDIO RANGE COMPRESSOR

Submitted by
Paul Leiffer
LeTourneau University

Some audio program material, such as classical music, has a large dynamic range. To avoid signal distortion during loud passages, recording is carried out at low levels. Unfortunately, noise in the recording and playback process then interferes with listening during the quietest passages. This problem can be partially overcome with an audio range compressor that provides more amplification during quiet passages than it does during loud ones.

Design such an audio compressor using operational amplifiers and passive circuit elements. The system should work for signals in the frequency range of 100 to 10,000 Hz. It should compress an 80-dB signal amplitude range by at least 6 dB.

8.28

DC MOTOR CONTROL

Submitted by
Paul Leiffer
LeTourneau University

Your client is developing a new line of super-fast model racing cars and needs a circuit to control a dc motor from an ac power source. The client wants to be able to control the placement of a current pulse that can be sent to the gate of an SCR by adjusting a single lever (accelerator). In other words, at each zero-crossing of the 12.6-V rms, 60-Hz power source, the circuit should generate a pulse whose delay is controllable from 2 ms up to as close as possible to 8.33 ms. Your circuit must be able to deliver a pulse of at least 5 V to an open circuit and 20 mA to a short circuit and be no longer than 1 ms duration. Specify the required dc power supplies.

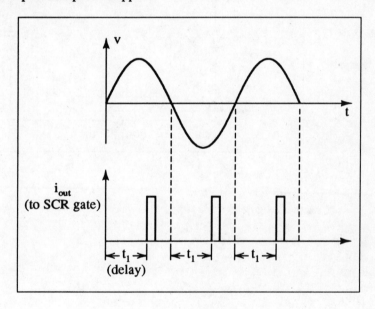

8.29

ELECTROMAGNETIC PULSE GUN

Your employer, a defense contractor, is developing an electromagnetic pulse gun to shoot down missiles. The prototype model of the gun works in part by charging a 1-F capacitor to 1000 V and then dumping the charge into a 1-Ω resistor by firing an SCR.

Develop the specifications for the SCR that should be used, particularly the di/dt, current rating, and voltage rating. The power loop has 10-μH equivalent series inductance.

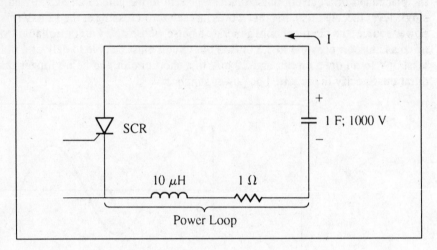

8.30

NEGATIVE VOLTAGE GENERATOR FOR LCD

Electronic System Design

Suggested by
Greg Clark
Research Applications, Inc.

Your company uses liquid crystal display (LCD) modules in several products. Some modules need a negative voltage for controlling the display contrast; other displays use a positive voltage for constrast variation. Because more than one type of display module might be used during the manufacture of a product, a voltage that can be adjusted between −4 V and +5 V must be provided on the board. (A potentiometer is adjusted until the contrast looks right to assembly personnel.) However, only a +5 V supply is provided on the board.

The circuit shown below can generate a negative voltage from the positive voltage supply. A square wave is produced by an oscillator made up of three 74HC04 inverters. The two diodes and capacitors provide the means of producing the negative voltage at the bottom end of the potentiometer. The output voltage from the potentiometer can be adjusted from the negative voltage to the positive supply voltage.

The maximum load on the adjustable voltage is 0.5 mA. Design the oscillator circuit so that the maximum ripple on the output is 0.25V in the worst case. To minimize power dissipation in the CMOS inverters, you should keep the frequency as low as possible.

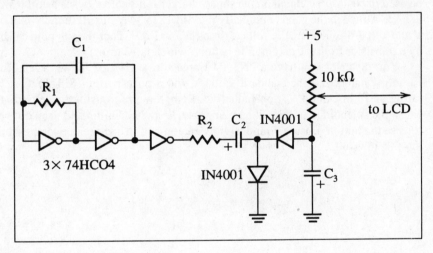

8.31

LED POWER INDICATOR

You are a design engineer for a toy manufacturer that is developing a new toy to take advantage of a current fad. The toy will use two MLED76 LEDs for the eyes, which will also serve as power-on indicators. Power from the four 1.5-V batteries (6 V) will be turned on by a switch.

Design the circuit to control the current through the LEDs. To ensure the eyes are bright enough, you need to keep the current in the LEDs above 30 mA. However, you also must guarantee that the specifications of the LEDs are not exceeded. The temperature range is 0° to 50°C. You must meet the performance requirements at minimum cost. If you can't meet the desired specifications over the entire temperature range, then predict the manufacturing yield at 25°C and over the full temperature range.

8.32

PUSH-BUTTON SOLENOID DRIVER

Transistor Switch

You own a gold and blue, super-charged Xenoflash convertible that will accelerate from 0 to 70 mph in 3.5 sec. Unfortunately, with the super-charger running, the fuel consumption decreases to 2.5 mpg and you can't afford the fuel costs. However, a solenoid can be used to control the clutch on the super-charger. This solenoid has a coil resistance of 280 Ω. Although the coil is nominally rated at 12 to 14 V, it will reliably pull in at 11.0 V and will drop out when the voltage drops below 1.0 V. Thus the solenoid can be driven by a transistor from the automobile battery, which is 12 to 14 V.

You discover you have a 2N3904 transistor, a 1N4007 diode, a handful of 5% carbon resistors (all the standard values), and a push-button (SPDT) switch with a rating of 10 mA at 24 V, non-inductive. Design a circuit using these parts that will cause the solenoid to pull in when you press the switch button and drop out when you release the button. Ensure your design is guaranteed to operate in a temperature range of −50° to +100°C.

8.33

SOLENOID DRIVER

Logic Interface

The solenoid control circuit you designed in Exercise 8.32 works well; however, you have found it difficult to maintain control of the vehicle while holding in the button that controls the clutch on the super-charger. Consequently, you have devised a logic circuit that senses the need for the super-charger and that will provide a high-level output when the solenoid is energized. The output of your logic circuit comes from a 54LS05 inverter. Nothing else is connected to this output.

Design an interface circuit between the 54LS05 and the solenoid so that the solenoid will be energized when the 54LS05 output is high and will drop out when the 54LS05 output is low. Again, ensure the circuit will work in the worst case over a temperature range of $-50°$ to $+100°C$. A 5-V power supply is available.

8.34

RESISTOR TOLERANCE TESTER

Your company has received several shipments of 220-Ω resistors that it suspects are not within the specified tolerance of $\pm5\%$. You are to make random tests on these resistors to verify the specifications.

Design a testing device that will produce a "go/no-go" test result. The test circuit configuration is a Wheatstone Bridge circuit, shown as follows:

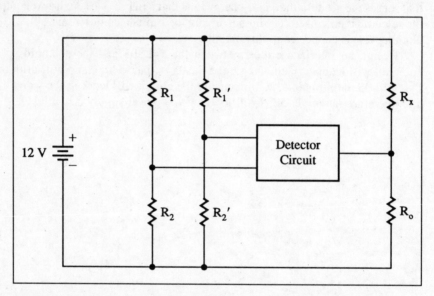

You determine that if you use 1% metal film resistors at nominal values in the circuit that will be used in a laboratory at constant 25°C, the measurement will be accurate enough.

Design the bridge and the detector for an automatic testing system using voltage comparators, logic gates, and any discrete components necessary to produce a "go/no-go" indication. The output indication should be a green LED for a good resistor and a red LED for a bad one. (You may assume green LEDs have the same electrical properties as red LEDs.) Assume +12-V and +5-V power supplies are available.

In addition to producing a complete electrical schematic of the design, write the operating instructions and determine the worst-case measurement performance.

8.35

MOTOR DRIVE

TTL Output Interface

Your employer is developing a line of toys for use by children traveling in a car. The toys will plug into the cigarette lighter for power, so they must operate on voltages from 12 to 14 V. One toy uses a motor for movement that has a resistance of 25 Ω, $\pm10\%$ and is controlled by logic, with the output coming from a 74LS00 gate. When the logic output is high, the voltage to be applied to the motor is to be at least 11 V; when the logic output is low, the voltage applied is to be less than 1 V.

Design the interface circuit that will convert the logic output to drive the motor. Because the same signal also must drive two additional 74LS00 gates, the circuit must not interfere with the existing logic level. Also, it must operate over a temperature range of 0° to 70°C without violating the specifications of any component. Assume a 5-V power source is available for the logic circuits. The best design would be the least expensive while meeting all other criteria.

8.36

WINDSHIELD WIPER CONTROL

TTL - FET Interface

The windshield wiper motor of your Xenoflash car is unusual: it's driven by a power MOSFET with its source grounded. You have just completed designing a revolutionary control system that will automatically adjust the wipe interval to maintain optimum visibility under any precipitation condition. Your control circuit uses a 54LS00 gate to give a logic high when the motor is to be turned on.

Design an interface circuit that will turn the MOSFET on and off, corresponding to high and low logic outputs, respectively. For it to do this, your interface circuit must generate a low voltage below 2.0 V and a high voltage above 10.0 V. Gate current is less than 0.5 mA. You may assume a 5-V power supply is available in addition to the car's battery (12-14 V). The temperature range is −55° to +100°C.

8.37

CMOS - TTL INTERFACE

The action-character toy your company is designing uses CMOS logic in a controller. The output of this controller comes from a 74HC08 gate that also drives two other 74HC00 gates inside the controller. In addition to the toy itself, an accessory—a chariot—is being developed for the toy to ride in. The chariot electronics input is made up of five 7400 gates that are to be driven from the same 74HC08 gate in the toy controller.

Design an interface between the 74HC08 and the five 7400 gate inputs. Because the chip limit for the toy has been reached, you must use discrete devices. The design must not violate any component specification over a temperature range of 0° to 70°C. Both the toy and the chariot use +5-V supplies. Keep costs to a minimum.

8.38

REVOLUTION COUNTER

Logic Interface

Your new Xenoflash convertible requires an oil change every 10,000,000 engine revolutions. Unfortunately, the car comes only with a tachometer. Because you spend a lot of time commuting in heavy traffic with the transmission in lower gears, you have no way of knowing how many revolutions the engine has endured. However, the camshaft is accessible and you can mount a Micro Switch in such a way that it will open and close once every two engine revolutions.

Design an interface between this switch and a 5400 logic gate so that the gate will see the open and closed positions as logic levels. This gate will be used to drive a counter that will count the engine revolutions. Note that an input to a gate should never be left floating because it is susceptible to noise. Also, the capacitance of the wire connected to the input of the gate can be as high as 200 μF, while the maximum engine speed will be 10,000 rpm. The switch has a maximum current rating of 10 mA. Assume both +5- and +12-V supplies with a common ground.

8.39

OPTICAL DETECTOR

Logic Interface

Design an optical revolution detector for the oil change notification system described in Exercise 8.38. Assume you can mount a slotted disc and a photo-emitter and photo-detector pair in such a way that the openings in the slotted disc may be counted. Design a circuit such that the photo-detector output will provide logic level signals to a 5400 logic gate.

8.40

DIGITAL SYSTEM DESIGN

LED Display

Design the logic, counters, and an LED indicator for the oil change notification system described in Exercise 8.38 so that once the circuit counts the 10 million revolutions, an LED will turn on. Design the system so that it can be manually reset when the oil is changed.

8.41

SECURITY SYSTEM SENSOR INTERFACE

RTL NOR Gate

You are a design engineer for Acme Solutions, Inc., which is developing a line of security equipment. One of the standard sensors is a simple SPST normally closed switch that has one pole grounded and the other pulled up to a +6-V supply by a 10-kΩ, 5% resistor.

Design a five-input RTL NOR gate that will interface with these sensor switches and will have a fanout of at least 6. You may use transistors and resistors only. The power supply is +6 V. The circuit must be guaranteed to work within your specifications over a temperature range of 0° to 70°C. Part of your design documentation should provide the operational and interface specifications. The best design would minimize the cost/noise-margin ratio. The noise margin is to be measured at worst case.

8.42

DTL NAND GATE

Xeno Motors is developing an electronic self-diagnostic system for its line of convertibles for commercial taxi service.

Design a three-input DTL NAND gate that can be standardized throughout the system. This gate is to be powered from the vehicle battery (12 to 14 V), must have a noise margin of at least 2 V, and must have a fanout of at least 15. Because the company has a policy against using integrated circuits, you must design this circuit using only discrete components. Your design documentation must provide specifications for the gate and the expected yield for manufacturing. You should keep average power dissipation below 40 mW at 14 V. The temperature range is −25° to +100°C.

8.43

DIGITAL TACHOMETER

Suggested by
Paul Leiffer
LeTourneau University

Design a digital tachometer for a four-cylinder, four-cycle engine. The input should be taken from the primary of the ignition coil, which will consist of 250-V pulses produced twice per engine revolution. The display should be seven-segment LEDs and should display engine speeds from 0 to 9990 rpm in increments of 10 rpm on a three-digit display. The display should update approximately twice per second. Assume a +5-V power supply is available.

8.44

WAVE SHAPER

Your company is developing a new line of digital cuckoo clocks that will be powered by a 6.3-V rms transformer from the 60-Hz line. Because the power lines have excellent long-term frequency stability, a clock running from the 60-Hz line should be accurate indefinitely (assuming no power interruptions). The project manager has determined that the new digital cuckoo clock will derive its frequency input from the power supply transformer.

Develop a rectangular wave clock pulse at 60 Hz from the 6.3-V rms sine wave. The output of your circuit should drive the clock input of a 74HC109. Assume a 5-V power supply is available.

8.45

ANALOG - DIGITAL INTERFACE I

The digital communications receiver your company is developing has a detector output, shown in the following figure, that drives into an open circuit. The detector output resistance is 1000 Ω.

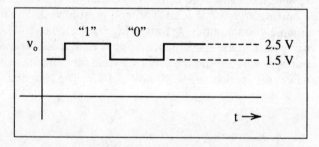

Design a circuit to convert this signal to a TTL compatible signal. Your circuit must be able to drive at least one 7400 gate. Assume a 5-V power supply is available.

8.46

ANALOG - DIGITAL INTERFACE II

The digital communications receiver your company is developing has a detector output, shown in the following figure, driving into an open circuit. The detector output resistance is 10,000 Ω.

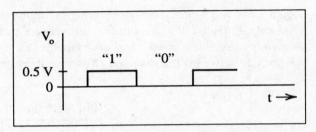

Design a circuit to convert this signal to a TTL compatible signal. Your circuit must be able to drive a 74HC00 gate. Assume a 5-V power supply is available.

8.47

DATA BUS DESIGN

Submitted by
Michael K. Williams
University of Massachusetts - Amherst

Your employer, a manufacturer of personal computers, is designing a new line of commuter computers using 54AS technology. The data bus for the computer has five 54AS00 gates connected to it as loads. To make the data bus structure accessible for expansion, allow one additional 54AS00 load or two 54LS00 loads. The data bus is to be driven by six 54LS05 gates, open collector inverters, on the main board. To provide for expansion, allow for one additional 5405 or two 54LS05 drivers.

Complete the design of the data bus by selecting an appropriate pull-up resistor. You must guarantee that your design will not violate the IC specifications over a temperature range of –25° to +100°C.

8.48

MEMORY SUBSYSTEM DESIGN

Submitted by
Manos Roumeliotis
City College of Thessaloniki

Design a 64k $\times$ 8 memory subsystem using the following 6264, 8k $\times$ 8 CMOS static RAM chip. Provide full address coding and produce a complete circuit schematic for the memory. Assume a power supply is available that meets your needs. Also assume WE and OE (active high) signals and 16 address lines are available from the CPU.

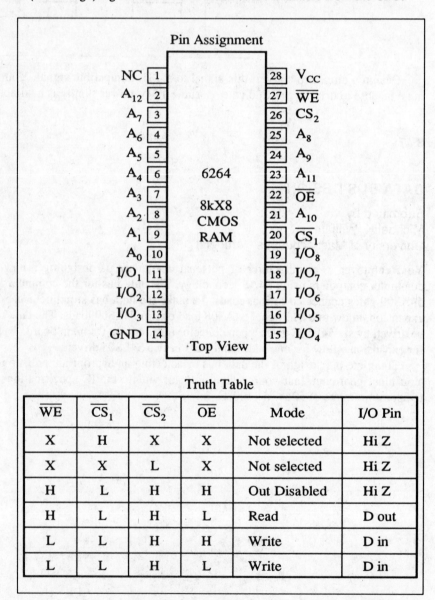

Pin Assignment

NC	1			28		V_{CC}	
A_{12}	2			27		$\overline{WE}$	
A_7	3			26		CS_2	
A_6	4			25		A_8	
A_5	5			24		A_9	
A_4	6	6264		23		A_{11}	
A_3	7			22		$\overline{OE}$	
A_2	8	8kX8 CMOS RAM		21		A_{10}	
A_1	9			20		$\overline{CS_1}$	
A_0	10			19		I/O_8	
I/O_1	11			18		I/O_7	
I/O_2	12			17		I/O_6	
I/O_3	13			16		I/O_5	
GND	14			15		I/O_4	

Top View

Truth Table

$\overline{WE}$	$\overline{CS_1}$	$\overline{CS_2}$	$\overline{OE}$	Mode	I/O Pin
X	H	X	X	Not selected	Hi Z
X	X	L	X	Not selected	Hi Z
H	L	H	H	Out Disabled	Hi Z
H	L	H	L	Read	D out
L	L	H	H	Write	D in
L	L	H	L	Write	D in

8.49

INTEGRATED CIRCUIT RESISTOR

Chip-level Design

Mountain Semiconductor, Inc., is developing an integrated circuit logic family. The substrate is p-type silicon. A resistor is to be made by diffusing in donors to make n-type material. The resulting n-type tub is insulated from the rest of the circuit by the p-n junction. The process line has several standard processes and allows only 0.5, 1.0, 1.5, 2.0, 2.5, and $3.0 \cdot 10^{16}$ donors/cm^3. Ohmic contacts will be made at each end, adding 0.02 mm to the length of the resistor.

Design a 1-kΩ resistor to be used in the logic family for minimum cost. Process constraints require that no surface dimension be less than 0.01 mm. Silicon surface area costs \$1/cm^2 and oven time costs \$100/hr $\cdot$ m^2. Time in the oven is proportional to the square of doping level ($t_0 = kN^2_D$), where 30 min is required for the lowest doping level. Diffusion depth is proportional to oven time (depth = 0.01 mm/hr). Ignore end and surface effects. Assume mobility of electrons remains constant at 475 cm^2/Vs at all doping levels. Specify the dimensions of the device and the doping level to be used.

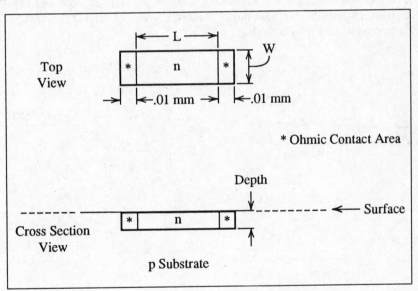

8.50

NMOS GATE DESIGN

Acme Semiconductor, Inc., is developing a line of NMOS integrated logic circuits using a depletion-mode NMOS pull-up.

Design an inverting buffer using the standard process with the following parameters. Use a 5-V power supply and specify the dimensions of the devices.
Standard process parameters:

Minimum feature size: 5 μm with 1 μm increments

$\mu_p = 230 \ cm^2/Vs$

$\mu_n = 580 \ cm^2/Vs$

$t_{ox} = 0.10 \ \mu m$

$\varepsilon_{ox} = 3.9\varepsilon_o \ (\varepsilon_o = 8.85 \cdot 10^{-14} \ F/cm)$

$N_A = 5 \cdot 10^{15} \ cm^{-3}$

$N_D = 5 \cdot 10^{18} \ cm^{-3}$

The threshold voltages are +1 and −1 volt for the enhancement and depletion mode transistors, respectively. This buffer must be capable of driving four 74LS00 loads and be compatible with 74HC00 logic.

8.51

CMOS GATE DESIGN

Mountaineer Logic, Inc., is developing a line of CMOS integrated logic circuits. You are to design a two-input NAND gate using the standard process with the following parameters.

MLI process parameters are:

Minimum feature size: 5 μm with 1 μm increments

$\mu_p = 230$ cm^2/Vs

$\mu_n = 580$ cm^2/Vs

$t_{ox} = 0.10$ μm

$\varepsilon_{ox} = 3.9\varepsilon_o$ ($\varepsilon_o = 8.85 \cdot 10^{-14}$ F/cm)

$N_A = 5 \cdot 10^{15}$ cm^{-3}

$N_D = 5 \cdot 10^{18}$ cm^{-3}

This gate must be capable of driving two 74LS00 loads. The design also must be capable of driving 10 identical gates. Use a 5-V power supply and specify the dimensions of the devices. Assume the threshold voltage for both p-and n-channel enhancement transistors is one volt.

Documentation for your design should include a specification sheet showing the four voltages and currents at the terminals for static conditions (V_{inLmax}, V_{inHmin}, I_{inL}, I_{inH}, V_{OL}, V_{OH}, I_{OL}, and I_{OH}) as well as power dissipation and propagation delays. Also plot V_0 vs. V_{in} and I_d vs. V_{in}.

OPTIONAL: Assume you want to produce a small-scale integrated circuit with the same pin configuration as the 74LS00. Lay out the four gates on a silicon chip including power and ground interconnections. Use minimum silicon area. Allow a minimum of 1 μm between features.

8.52

ANALOG ELECTRONICS - RELIABILITY

Submitted by
Joseph H. Wujek
San Jose State University

The *Arrhenius model*[1] is a popular model used in reliability physics for estimating failure-rate dependence with temperature. In this model, failure rate $\lambda(T)$ is given by the following equation:

$$\lambda(T) = Ce^{[-\phi/kT]},$$

where C is a constant, ϕ is the activation energy, T is the absolute temperature, and k is Boltzmann's constant ($8.62 \cdot 10^{-5}$ eV/K°). When the model is applied to semiconductors, T is usually taken as the device junction temperature, T_J.

Design a test-fixture for transistors. A prototype silicon NPN RF-power transistor is to be used in a power transmitter. Physicists have determined that the activation energy for thermal failure-modes is 0.38 eV. The chips should be *burned-in*, that is, the devices will be operated at a junction temperature higher than the use (that is, field) condition so as to cull the majority of early failures. The burn-in board to be used with the device should stress the part to a junction temperature of 175°C. The devices are to be burned-in for 168 hr (1 wk).

Relevant specifications and use conditions follow. Note that in field use, a heat sink is employed.

Table 8.52. Use Conditions.

Characteristic	Value	Symbol	Remarks
Collector-emitter voltage	10 V	V_{CEO}	rms value
Collector current	0.46 A	I_C	rms value
Thermal resistance, case to heat sink	0.6°C/watt, max	θ_{CS}	
Thermal resistance, heat sink to ambient	3.0°C/watt, max	θ_{SA}	
Ambient temperature	35°C	T_A	

---

[1]Svante August Arrhenius (1859-1927) was a Swedish physical chemist perhaps best-known for his theory of electrolytic dissociation, for which he received the Nobel Prize for chemistry in 1903.

Table 8.52. Device Specifications.

Characteristic	Value	Symbol	Remarks
Collector-emitter voltage	40 V, max	V_{CEO}	
Collector current	3.0 A, max	I_{Cmax}	
Thermal resistance, junction to case	4.0°C/watt, max	θ_{JC}	
Current gain	15 min, 45 max	h_{FE}	$0.15 \leq I_C \leq 1.5A$ $V_{CE} = 10$ V
Current gain	10 min	h_{FE}	$1.5 \leq I_C \leq 3.0A$ $V_{CE} = 10$ V
Thermal resistance, junction ambient	4.0°C/watt, max	θ_{JA}	
Junction temperature	175°C/max	T_{Jmax}	

You are to do the following:

1. For the use conditions, compute the rms power, the junction temperature, the case temperature, and the heat sink temperature.
2. Assuming the failure rate is invariant in time for constant temperature, how many hours of field-use corresponds to the 168 hr of burn-in? What fraction will survive to this time?
3. Design a bias network for the transistor in the burn-in board. Assume the device will be burned-in under dc conditions with an ambient temperature of 25°C ±2°C. The collector-emitter voltage should be 10 V. The network should maintain the desired junction temperature of 175°C with a tolerance of +0, –10°C for any value of h_{FE} within the specified limits. One power supply is to be used and is adjustable over a range of 5 to 100 V, with a regulation of ±0.5% over 0 to 25 A. The burn-in board should have 20 positions for devices; five such boards will be fabricated. Additional power supplies can be obtained as needed, but no more than one supply per board may be used.
4. Suppose the activation energy is 0.46 eV instead of 0.38 eV. Rework (2) with this new value of ϕ and compare the results to the earlier result.

8.53

DIGITAL LOGIC PROBE

Suggested by
Michael E. Parten
Texas Tech University

Design a digital logic probe with three LED indicators to indicate HIGH logic levels, LOW logic levels, and PULSE signals for TTL logic. For floating inputs, both HIGH and LOW lamps should be off. The PULSE LED should be activated for approximately one-third second on either the rising or the falling pulse transition. Pulse trains should cause on/off flashing at a steady rate.

Suggest additional specifications to make this system more attractive or useful than a similar product. Include a cost estimate of the system based on the cost of parts for your prototype.

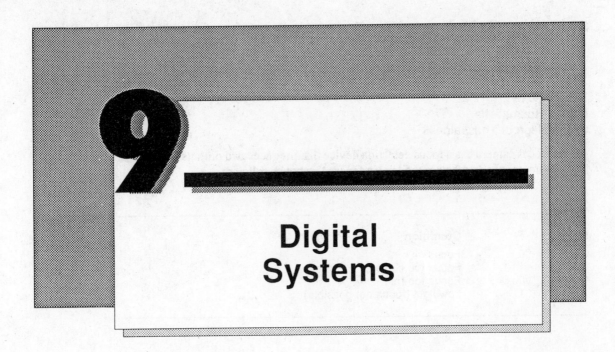

9

Digital Systems

9.01

DIGITAL BALLOT BOX

Submitted by
Manos Roumeliotis
City College of Thessaloniki

A company needs a digital logic system to help with its secret voting procedure at board meetings. The board consists of six members; each is to have a push-button switch attached under the table. If the member presses the button (logic 1), the vote is yes; otherwise, the vote is no (logic 0). A YES/NO display should give the result of the voting. The YES sign lights up if the majority of votes favors the motion; otherwise, the NO sign lights up. The board president's vote counts as three votes, the two vice presidents' votes count as two each, and the votes of the other three members count as one each. In case of a tie, the president's vote decides the outcome.

Design the logic for this voting procedure using approved gates. Provide a logic 1 for the YES indicator and a logic 0 for the NO indicator.

9.02

CAMERA AUTOFOCUS DESIGN

Submitted by
Manos Roumeliotis
City College of Thessaloniki

An autofocus camera has a focus detection device that produces two outputs, F_1 and F_2. These outputs specify whether the lens is focused or not as follows:

F_1	F_2	Condition
0	0	Focus ok
0	1	Focus too close
1	0	Focus too far
1	1	Low light (focus not possible)

The logic circuit of the camera accepts inputs and produces pulses for the micromotors that focus the lens. A pulse train in line Z_1 turns the lens barrel clockwise (shortening the focus), while a pulse train in line Z_2 turns it counterclockwise (lengthening the focus). All of these operations, of course, take place if the shutter release button is pressed. When the lens is focused, the shutter is released. If the light is too dim for the autofocus system to operate, the shutter is locked.

Design the logic circuit that performs these operations. Provide three outputs, Z_1, Z_2, and Z_3 (shutter release) that are TTL compatible. The pulses on lines Z_1 and Z_2 should occur once every millisecond with a 50% duty cycle until focus is correct. The inputs F_1, F_2, and F_3 (shutter release button) are also TTL compatible. Assume +5-V dc power supply, as well as 1-kHz square wave, P is available. Design for minimum number of approved integrated circuit modules.

9.03

SYSTEM VARIABLE DECODE CIRCUIT

Suggested by
Jeff Reel
HRB Systems

STLX, a steel producing plant, recently acquired a new pickler line that uses a combination of chemicals to clean the cast steel. Three cylinders are used to store and disperse the solution to the line, and as a safety precaution, the temperature and pressure are monitored in each. If the temperature and pressure fall below 80°C and 2 atm, respectively, control signals to an alarm will be generated. A control circuit is to generate an alarm if the temperature and the pressure fall below the desired level in at least one of the cylinders or if the temperature or pressure falls below the desired level in all the cylinders. There are two outputs for each cylinder i: $P_i = 1$ if pressure is above threshold and $T_i = 1$ if the temperature is above threshold.

Design the logic circuit that decodes the six system variables and generates this alarm signal. (Do not design the sensing circuits or alarm circuit.) Minimize the cost using approved digital integrated circuits. The ambient temperature may go as high as 100°C.

9.04

COMBINATION LOCK

Suggested by
Paul Leiffer
LeTourneau University

Problems have arisen with the monitoring system for STLX's pickler line (see Exercise 9.03). The sensing circuits somehow became "tweaked." Your manager wants to put a 10-bit binary combination lock on the panel that contains the circuitry. Using only approved TTL integrated circuits, design a combination lock that will light an LED when the proper pattern of 1's and 0's is set on the bank of 10 switches. Include a method by which the lock can be easily reprogrammed to recognize any desired pattern. A logic 0 should turn on the LED.

9.05

LOGIC DESIGN

Your neighbor works to support four teenagers, A, B, C, and D. The kids are alone in the apartment from the time school is out until 5:30 P.M. To keep peace in the family, the teenagers have agreed to the following rules:

1. They will confine themselves to three rooms in the apartment: the living room, the bathroom, and the kitchen.
2. At least one person will always be in the living room to prevent the cat from eating the goldfish.
3. No one will be alone in the kitchen.
4. Not more than one person will be allowed in the bathroom at any time.
5. All four will not be allowed in the same room at the same time.
6. A and C can't be in the same room alone.
7. B and D can't be in the kitchen alone.

Your neighbor, an electrical engineer, has developed a sensing device and placed one in each room. This device can identify each of the four teenagers and has four outputs—a, b, c, and d. It produces a 1 on the appropriate output when the corresponding individual is in the room; for example, $a_k = 1$ when A is in the kitchen.

Design a logic system that will produce a logic signal when an illegal occupancy combination occurs in any room. If an illegal combination does occur, a bell should chime; the teenagers have agreed in such a case to move to a legal condition. This bell should be placed in a strategic location so it can be heard throughout the apartment. A logic 0 is required to make the bell chime. The best design would have the lowest total chip cost using approved integrated logic circuits. The temperature range is 10° to 35°C.

9.06

VLSI IC FULL ADDER DESIGN

Submitted by
Manos Roumeliotis
City College of Thessaloniki

You are part of a design team that is designing a special-purpose military VLSI integrated circuit. It is to be CMOS, and the design requires that it contain several full adders. To minimize the chip area, the area occupied by a full adder should be minimized. Only two-input gates are to be used. The chip area occupied by various gates is given in the following table:

Gate	Chip Area (in μm^2)
AND	600
OR	600
NAND	400
NOR	400
XOR	1600
Inverter (1 input)	200

Design a full adder that occupies the minimum chip area. Allowance for wiring is included in the table.

9.07

OVERFLOW DETECTOR

Submitted by
Manos Roumeliotis
City College of Thessaloniki

The 2's complement representation is used to store both positive and negative numbers in a computer. The ALU (Arithmetic & Logic Unit) of the computer is able to add the numbers in two eight-bit registers A and B and put the result in the eight-bit register C, again in 2's complement representation.

Using approved gates, design the logic that gives an output of 1 whenever an overflow occurs during the addition of the numbers in A and B.

9.08

ALU DESIGN

Submitted by
Manos Roumeliotis
City College of Thessaloniki

The ALU of a microcomputer has two control input lines, S_0 and S_1, and a carry input line, C_{in}. Three registers are available to the ALU: the accumulator (AC) and the data registers R_1 and R_2. Depending on the inputs, the ALU should perform the following functions:

S_0	S_1	$C_{in} = 0$	$C_{in} = 1$
0	0	$AC \leftarrow R_1 + R_2$	$AC \leftarrow R_1 + \bar{R}_2 + 1$
0	1	$AC \leftarrow R_1 + 1$	$AC \leftarrow \bar{R}_1 + 1$
1	0	$AC \leftarrow R_1 \wedge R_2$	$AC \leftarrow R_1 \vee R_2$
1	1	$AC \leftarrow R_1 + R_1$	$AC \leftarrow R_2 + R_2$

+ denotes addition, $\leftarrow$ denotes register transfer, $\cup$ denotes logical AND, $\cap$ denotes logical OR, and $\bar{X}$ denotes complement. Using approved digital integrated circuits, design an ALU to perform these functions. Show your design for two data bits, d_0 and d_1.

9.09

BINARY SUBTRACTOR

Suggested by
Michael E. Parten
Texas Tech University

Design and simulate a five-bit binary subtractor that can subtract both positive and negative numbers. Provide toggle switches to set the two input operands and LEDs for the output. There should be a single push button for performing a subtraction. When the button is pushed, the output register should be cleared; when it is released, the new value should appear.

9.10

BCD COUNTER

Submitted by
Manos Roumeliotis
City College of Thessaloniki

SPEE DEE Mfg., Inc., has been investigating plant costs and expenditures over the past year for the purposes of cutting costs and increasing productivity. The industrial engineer in charge of the project has written a real-time line simulation program using data that has been compiled over time. Through her simulation, she has found discrepancies in the data given to her involving the through-put time of products on an assembly line. She requests that you place timers at various points on the line that are to be triggered on and off by "electric eyes" placed across the path of the products on the line. A timer is to reset if a product passes or if 1 minute has elapsed, whichever occurs first. A 60-Hz clock signal and signals from the "electric eyes" with TTL level outputs are provided. Because of budget constraints, you must optimize for a minimum number of integrated circuit chips and use only approved parts.

9.11

ERROR DETECTING CODE

Suggested by
Manos Roumeliotis
City College of Thessaloniki

Selene, Inc., manufactures high purity crystals from lunar dust. Every day the operators transmit production information to the home office, where you work. Recently, sun spot activity has been interfering with the communication link. Several errors have been made in interpreting the data that have cost the company considerable money in the futures market.

Devise a code that will allow the receiver to detect errors in transmission. Because the error rate is not large, single-bit error detection is deemed satisfactory. For single-bit error detection, each code word must be different from every other code word by at least two bits. Design the most efficient code that will allow the coding of decimal information, with codes for the decimal digits 0–9.

9.12

EGG COUNTER I

Gray Code

Submitted by
Manos Roumeliotis
City College of Thessaloniki

You are an electrical engineer in charge of packaging technology for a food products company. The egg packing line utilizes a machine that inserts eggs into containers, each holding one dozen eggs. To further automate the packaging process, you want to develop a system to count the eggs as they are packed. To reduce the possibility of error, you decide the eggs need to be counted in a Gray Code.

Design a modulo-12 Gray Code counter.

9.13

EGG COUNTER II

Sequential Circuits

Suggested by
Manos Roumeliotis
City College of Thessaloniki

The technical oversight committee has authorized you to design a modulo 12 Gray Code counter/controller for its new automated egg packing machine. The packing machine produces a 1-ms TTL level pulse upon successful insertion of an egg into the container.

Design a circuit so that when the twelfth egg is inserted, the circuit will produce a 1-ms TTL level pulse on its output, thus causing the machine to eject the full container and prepare an empty one.

Further, for safety reasons, you need to include one other feature in your circuit. One time, a worker was injured when the power supply was momentarily interrupted and then came back on, causing the logic system to go into an unused state and the machine to become erratic. Therefore do not allow unused states to persist; that is, the next state must be a valid count state.

9.14

VARIABLE RING COUNTER

Suggested by
Michael E. Parten
Texas Tech University

Design a four-bit variable-length ring counter. The circuit should have two two-position (SPDT) switches to select a length of one, two, three, or four bits for the register; for any length, the unused bits should always be 0. The circuit should be self-starting and self-correcting for any length sequence. Assume an appropriate power supply and clock signal are available.

9.15

SELF-CORRECTING RING COUNTER

Suggested by
Michael E. Parten
Texas Tech University

Design a self-starting and self-correcting eight-bit ring counter with a circulating 0. All eight bits should be shown on an LED display. The counter should automatically return to a valid state in a few cycles regardless of its starting state and without a reset button.

9.16

SERIAL TO PARALLEL DATA CONVERTER

Submitted by
Jeff Reel
HRB Systems

A synchronous serial signal and its clock are being received by a system. The serial signal is a continuous bit stream. Before the signal can be processed by the system, it must be converted into a parallel format. The serial data is formatted into eight-bit words, of which the first three bits make up a synchronization pattern. The serial clock frequency is 5 MHz. The most significant data bit is received first. The synchronization pattern is 111 and is guaranteed not to show up in the data stream.

Design a circuit using approved integrated circuits that will receive and convert the serial data to parallel data, with the words aligned on the proper boundaries. The circuit also should produce signals that indicate when the system can begin processing the converted data (data is valid) and when the system can expect the parallel words (clock).

9.17

CYCLOPS BRAKE LIGHT MODULE

A single additional brake light centered high on the car (a cyclops light) has been shown to reduce the likelihood of a rear-end collision. Design a module to enable a user to add such a light to a car that doesn't have one.

The module must work on a car that uses the same light bulbs for brakes and for rear turn signals. (In these cars, enabling the turn signal in a particular direction disables the brake light on that side.) To provide inputs to the module, you can connect to the wire going to the right brake/turn lamp, the left brake/turn lamp, and ground. Once connected, your module should light the cyclops light when the brakes are applied but not change the action of the original brake/turn lights.

Provide a complete logic diagram using approved digital integrated circuits. Assume incoming signals are logic compatible (1 = light on) and a proper power supply is available. The output to the cyclops light should be a logic 1 when the light is to be on. (You do not need to design the high-current drive to actually power the light.)

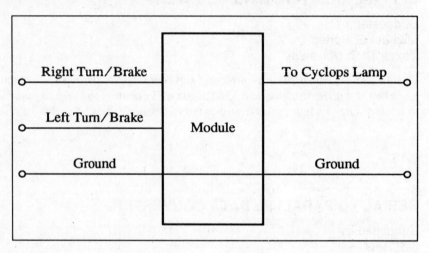

9.18

VENDING MACHINE DISPLAY

Submitted by
Manos Roumeliotis
City College of Thessaloniki

A vending machine indicator has a two-digit, seven-segment display to indicate the amount of money put in the slot before a selection is made. The range of the display is 00–95¢. Any combination of dimes, nickels, or quarters is accepted. Assume circuitry is available that has an output line for each type of coin and that produces a positive logic level pulse on the appropriate line when a coin is deposited.

Design a digital circuit using approved digital IC chips that will count the amount of money deposited. The total value of money deposited is to be stored in a two-digit register in BCD format. Design the logic and the register only and provide a complete circuit schematic and wiring diagram. Assume an appropriate power supply is available.

9.19

TRAFFIC LIGHT CONTROLLER

Submitted by
Manos Roumeliotis
City College of Thessaloniki

Design the logic for a traffic light controller. Assume a clock signal is available whose frequency is 0.5 Hz. The sequence of events is to be as follows:

(NS R, EW G) for 16 sec
(NS R, EW Y) for 2 sec
(NS G, EW R) for 10 sec
(NS Y, EW R) for 2 sec

The sequence is then repeated. If the circuit fails, all lights should be red.

Use digital integrated circuits and assume you have an appropriate power supply available for the controller logic. The output of your controller should have six outputs: NS–G, R, Y and EW–G, R, Y. A logic high will be used to energize the lamp. (You don't need to design the high-current driver to actually power the lamps.)

9.20

SEQUENTIAL TURN SIGNALS

Suggested by
Michael E. Parten
Texas Tech University

Design a circuit that controls six taillights on a car (three on each side). Use toggle switches for the left turn, right turn, and emergency flasher. For a right turn, the three right-hand lamps should be activated and the left-hand lamps should be off. The three ON lamps should follow the four-step sequence: all off; inside light on, other two off; two lights on, outside light off; all three lights on; and all lights off. The cycle should repeat about once a second. Operation of the left-hand light is the same. When the emergency flasher is enabled, all six lights should flash on and off in unison, with a frequency of about 1/2 Hz.

Use a push button to represent the brake pedal switch. When the brakes are on, all six lights should be on continuously, unless the right- or left-turn signal is on, in which case, the three taillights for the turn signal should operate normally and the other three should be on continuously. Assume an appropriate power supply is available and design the digital logic and sequential circuits only.

9.21

ELEVATOR CONTROLLER

Suggested by
Michael E. Parten
Texas Tech University

Design a system to control an elevator in a seven-story building. Your system should respond to floor-call switches and in-car floor-select switches and determine where to go next. All signals are TTL compatible.

9.22

BINARY MULTIPLIER

Suggested by
Michael E. Parten
Texas Tech University

Design and simulate a binary multiplier for two four-bit unsigned operands using the conventional "add-and-shift" algorithm. The circuit should produce an eight-bit product after four steps. Provide eight toggle switches for the two input operands and eight LEDs for the output.

Include a single push button for performing a multiplication. When the button is pushed, the product register should be cleared; when the button is released, the new value should appear.

Assume an appropriate power supply and clock signal are supplied.

9.23

FINITE SEQUENCE PULSE GENERATOR

Suggested by
Michael E. Parten
Texas Tech University

Design a system with the following specifications:

- Eight toggle switches for setting a binary number N
- A free-running clock input that can be connected to a pulse generator (assume the pulse generator produces positive 5-V pulses at 1 MHz)
- A start push button
- A single, normally high output

When the start button is pushed, exactly N clock pulses should be passed to the output. The start signal is not synchronized with the clock and can change at any time. However, no shortened pulses should appear at the output. The start signal should be at least one clock pulse in length.

9.24

FREQUENCY COUNTER: JOB INTERVIEW

Suggested by
Michael E. Parten
Texas Tech University

You are interviewing for a highly competitive summer position at an outstanding instrument manufacturer. There are 200 students applying for 10 positions, so the recruiter is asking technical questions as part of the interview process. She asks you to design a "floating-point" frequency counter.

Use a 1-sec reference period and measure frequencies up to 1MHz. Display only the two most significant digits of the frequency and a multiplier. If C is the most significant digit, D the next, and E the multiplier (power of 10), then the following frequencies would be indicated by the appropriate values of C, D, and E:

Frequency	C D E
99 Hz	9 9 0
100 Hz	1 0 1
1 kHz	1 0 2
etc.	

The system should include an input for the signal to be measured and a push button to perform the measurement.

Suggest additional specifications to make this system more attractive or useful than a similar product. Include a cost estimate of the system based on the cost of parts for your prototype.

9.25

TV VIDEO GAME

Submitted by
Paul Leiffer
LeTourneau University

A new video game uses a brightened line on a standard TV screen. Develop digital logic that identifies a single, specific TV line by counting individual horizontal sync pulses (which begin a line) following a vertical synchronization pulse (which begins a new frame).

Assume you have circuits that detect these two types of synchronization pulses and provide logic-level signals, S_H and S_V.

9.26

CHANGE DISPENSER

Submitted by
Dennis G. Smith
The University of Alabama at Birmingham

Design a combinational digital logic-based change dispenser for a fast-food restaurant. Change must be guaranteed to consist of a minimum number of coins. For example, 76¢ change should consist of three quarters and one penny, not seven dimes and six pennies. Change is held in four stacks, one each for quarters, dimes, nickels, and pennies. Each stack has one or more coin dispensing actuators; for example, the quarters stack would have three actuators, one each for one, two, and three quarters. Each stack has an out-of-change sensor associated with the largest actuator; for example, the quarters' sensor would sense when there were fewer than three quarters in the stack. Assertion of any of the out-of-change sensors should cause a light to light up and prevent the actuators from operating. The user input device consists of 99 keys arranged in a square but with no key for zero, that is, purchases of an even dollar amount. The keys' output are individually available and are not prewired in any particular configuration. The keys are labeled with the cents total of the purchase amount; for example, for a purchase of $4.24, the user would push the key labeled 24, which would cause 76¢ change to be returned. Assume a suitable power supply is available.

9.27

STEPPER MOTOR DRIVE FOR AN ELECTRIC NAIL CLIPPER

Submitted by
Paul Leiffer
LeTourneau University

Your product group is developing an electric nail clipper. Part of the hardware involves a stepper motor control.

Develop a sequential circuit to drive a basic four-phase stepper motor. The required four output signals have a 50% duty cycle and overlap 50% with each adjacent step signal, shown as follows. A variable frequency clock generator is available and all signals are TTL compatible.

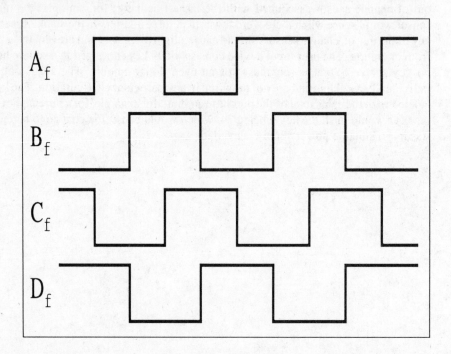

9.27a.
Make the motor drive system reversible using a direction signal: $F = 1$ for clockwise rotation as shown in the timing diagram above and $F = 0$ for counterclockwise rotation with the order of the signals reversed.

9.28

STEPPER MOTOR DRIVE CONTROL

Suggested by
Michael E. Parten
Texas Tech University

Design a circuit to control the drive circuit for a stepping motor. (See Exercise 9.27 for the required drive signals.)

A 16-bit control word is received from a computer bus. The word structure is MSB (bit 15) = mode bit; bit 14 = direction bit, 1 = ccw rotation, 0 = cw rotation; bit 13 - bit 0 = number of steps in binary.

- If the mode bit = 1, the system controller is to step (at 100 steps/sec) the stepping motor continuously in the CCW direction until a mechanical limit switch is activated, then stop for 0.1 sec, and finally, step the motor at the same rate in the CW direction until the exact number of steps specified is completed.

- If the mode bit = 0, the system controller is to step the stepping motor from its present position in the direction specified by the direction bit until either limit switch has been reached or the step number is completed, after which it should stop.

You have a 400-Hz TTL compatible clock available along with TTL compatible limit signals.

9.29

DIGITAL CAPACITANCE DEVICE

Submitted by
Paul Leiffer
LeTourneau University

Recent studies by your firm show a potential market for a simple, inexpensive capacitance measuring device. A digital system with LED digital readout is required.

Design a device capable of measuring values of capacitance from 100 pF to 1 µF. The readout should provide two digits for capacitance value plus two more digits indicating the multiplier.

9.30

SUCCESSIVE-APPROXIMATION A/D

Suggested by
Michael E. Parten
Texas Tech University

Design an eight-bit successive-approximation analog-to-digital converter. The analog input signal is from −5 to +5 V. Also determine the maximum sample rate allowed on the input for your design.

9.31

DOUBLE-RAMP A/D

Suggested by
Michael E. Parten
Texas Tech University

Design an eight-bit double-ramp type analog-to-digital converter. The analog input signal is from −5 to +5 V. Also determine the maximum sample rate allowed on the input for your design.

10

Electromagnetics

10.01

CAPACITOR DESIGN I

AstroNav, Inc., a major manufacturer of interplanetary inertial guidance systems, has placed an order for 100,000 0.024-μF, 75-V, mylar capacitors designed for minimum PC board footprint with a maximum height of 1 cm.

Make a drawing that shows how the capacitors are to be assembled as well as what the original mylar and foil sheets look like before assembly, including all dimensions. Develop a specification sheet showing electrical and mechanical specifications.

Aluminum foil: 0.05 mils (0.05/1000 in.)

Mylar sheet: Relative dielectric constant = 3.11

                          Dielectric strength = 18.1 MV/m

                          Available thicknesses = 0.1, 0.2, 0.3, . . . mils

10.02

CAPACITOR DESIGN II

Your company has received an order for 500,000 capacitors that must meet the following specifications:

Capacitance:	0.013 µF, (−0%, +28% tolerance)
Working voltage:	45 V minimum
Temperature range:	−40°F to +200°F
Frequency range:	1 to 10 MHz

Design the capacitors using aluminum foil and polyester (Mylar), the specifications of which follow:

Available thicknesses:	0.1, 0.2, 0.3, . . . mils, @ 25°C (thousandths of an inch),
	±10% manufacturing tolerance
Dielectric constant:	3.11 @ 1.0 MHz
	3.04 @ 10.0 MHz
Dielectic strength	18.1 MV/m
Thermal coefficient of expansion:	≤ ±9.5 PPM/°C

Aluminum foil is available in a thickness of 0.2 mils, negligible thermal expansion.

Also make a drawing that shows the assembly of the foil and dielectric sheets including dimensions.

For the customer, develop a specification sheet showing electrical and mechanical specifications.

10.03

DIELECTRIC VOLTMETER

Using the general structure shown in the following figure, design an electrostatic voltmeter that will have a displacement of 2 mm/kV at 100 kV and work for a range of 0 to 250 kV. The voltmeter utilizes the force tending to draw a plate of dielectric into the space between the parallel plates of a capacitor. The force on the dielectric plate is balanced by a spring so that the plate's position is a measure of the voltage. Be sure to specify all dimensions and the spring constant.

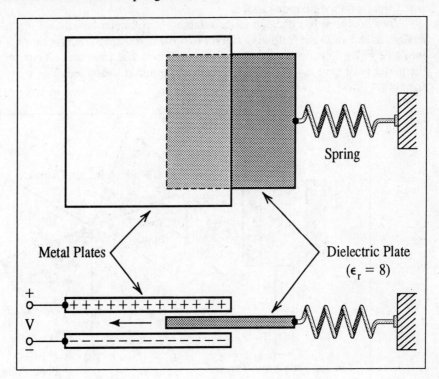

10.04

KELVIN'S ELECTROSTATIC VOLTMETER

Lord Kelvin was one of the early pioneers in electromagnetics. He was also an exceptional instrument maker. One instrument he made was an electrostatic voltmeter that consisted of fixed and moveable pie-shaped metal plates that could pass close to each other. The system of plates was perfectly balanced so that any voltage applied between the fixed and moveable plates would cause the plates to rotate into alignment with the fixed plates (thus maximizing system capacitance). This movement was restrained by a small weight on a rigid radial spoke.

Your company has decided to manufacture a voltmeter based on Lord Kelvin's design, as shown in the following figure. However, you are to redesign it so it will fit into a 12 • 12 • 5 cm enclosure. The vanes are to be made from brass 2 mm thick, and the meter must give a 45° clockwise rotation for 1,000 V applied. You may neglect fringing.

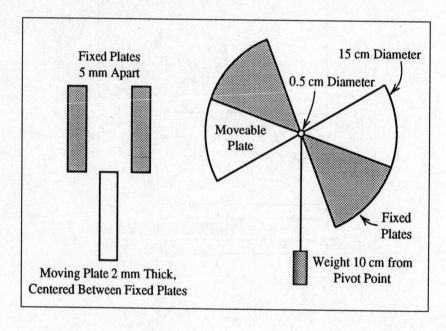

10.05

ELECTRICAL ACCIDENT INVESTIGATION

As a registered professional engineer, you have been asked to testify in a court case involving alleged injury to a child caused by an electrical current. This current was produced by a toy, part of which is shown in the following figure:

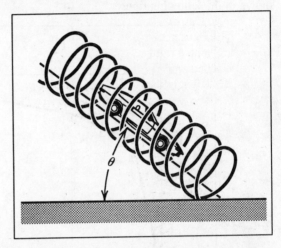

A toy car, which carries a magnet, is placed on an inclined plane of angle θ, whereupon it rolls down the hill through a tunnel, which is really a long coil. The movement of the magnet through the coil induces a current in the coil that actuates a relay. It is this induced current you are to investigate.

After running a series of experiments you accumulate the following data:

$V = \alpha v$, where V = voltage induced in the coil

v = velocity of the car

$\alpha = 15$ (V sec/cm)

$F = -i\beta$, where F = force tending to slow the car

i = current in the coil

$\beta = 20$ (g cm/sec^2 A)

The induction coil and relay coil in series have a total inductance of 1 H and a resistance of 1 Ω. The rolling friction of the car is negligible. The mass of the car is 10^3 g. The slope is 30°. The car begins at rest and continues until it hits a stop, which it does in approximately 5 sec. The experimental equations hold throughout its journey.

What is the maximum current that flows in the coil and when does it occur? You need to explain to the jury whether it would be possible to redesign the toy to reduce the maximum current to one-fifth its present value without significantly changing the motion of the car.

10.06

ELECTROMAGNET

Design a magnet that produces a flux density of 1.5 webers/m^2 in an air gap that is circular in cross section (1 cm diameter) and 2 mm long. You are to use a new magnetic material that can be cast or machined to any shape and that has the magnetic properties shown in the following figure:

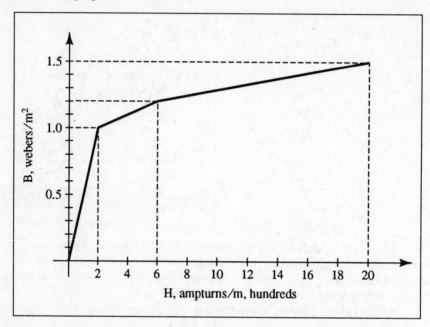

A 12-V dc source is available to power the magnet. Your design should show all dimensions of the core structure and specify the size wire and number of turns to be wound on the core. Also indicate how much current the magnet will draw from the source.

10.07

ELECTROMAGNETIC LATCH

Reducing loss of life in the event of a fire in a high-rise building is highly dependent on preventing toxic smoke from spreading throughout the building. This is done by isolating the fire areas with doors and ventilation shutters.

Design an electromagnetic latch to hold a fail-safe emergency smoke shutter. As long as power is supplied from a 1-A dc current source, the latch should hold up a mass of 15 kg. If power fails, the mass and the bottom core piece should be allowed to fall. The basic design is as follows:

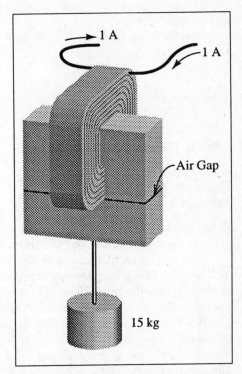

A U-shaped structure and a straight bar are made from square steel bar stock whose permeability is 5000 μ_0. Because of surface irregularities and corrosion, the air gap where the straight bar joins the inverted U can be no less than 0.1 mm. To wind the coil, you can use #28 copper magnet wire, which with its insulation, is approximately 0.5 mm diameter.

The magnetic attraction available to hold the weight can be increased by using material with a larger cross-sectional area, by increasing the number of turns of wire in the coil, or by making the magnetic path length shorter. The number of turns is limited by the size of hole (window) in the structure through which all the turns must pass.

Choose dimensions for the magnetic structure so that the weight will be supported but the volume of steel used in the completed device will be minimized. Also, indicate the number of turns in the coil.

10.08

EARTHWORM RETRIEVER

After slaving through four years of college, including two summer sessions, without a break, you decide to take a week to go fishing. After flying into a remote fishing camp, your party discovers you forgot to bring any bait. You have heard that by passing electrical current through the earth you can "tingle" the worms enough to cause them to come to the surface, where you can harvest them. You have a small portable 110-V ac generator available that is fused at 5 A, so you must not exceed this limit. You want the current to be as large as possible, however, so you get a good harvest. A book on earth resistivity you found in the cabin says that the soil in the vicinity has 1000 to 1500 Ω/cm^3 resistivity.

Determine the diameter, length, and spacing for two metal rods to be driven into the ground and connected to the portable generator to "raise" worms.

10.09

FM FILTER DESIGN

Submitted by
Dennis P. Yost
Syncro Development Corp.

A radio station radiates such a strong signal at 101.9 MHz that other nearby radio stations, such as FM102.5, can't be tuned in on an analog tuner.

Design a notch filter that will trap 101.9 MHz but allow your stereo to receive 102.5 MHz. The indoor antenna feeder line and the input impedance of the stereo antenna terminals are both 300Ω. You need a minimum attenuation difference between 102.5 and 101.9 MHz of at least 10 dB.

Provide a schematic of the filter, an electrical specification sheet (including a plot of the filter's frequency response), and working drawings for the construction of the necessary inductor. Specify the materials to be used, dimensions of the components, and construction techniques for winding the inductor.

10.10

TUNING STUB

A toy company is designing a magnetically levitated model train that will be controlled with a radio at 27 MHz. The input impedance to the radio is $100 + j100\ \Omega$ but with a tolerance of $\pm20\%$. This radio is fed by a 300-Ω $\pm10\%$, twin-lead transmission line. The radio is sent through a short run on the production line, whereupon you learn that the SWR may be as high as 3.5; for reliable operation, the SWR should not be greater than 1.5. You can remedy the problem by using a short section of the same transmission line as a stub tuner.

Design a process that can be implemented on the production line so that the technician can make one measurement and cut the stub to a length that will guarantee an adequate SWR. Be sure to specify the tolerance with which the stub must be cut and the position and tolerance of the placement of the stub.

10.11

RF TRAP

Submitted by
J. F. Corum
Battelle Memorial Institute

You tuned your FM radio to receive a coded FM message on a very low-power, clandestine channel operating at 102.5 MHz. Unfortunately, a local rock-and-roll station at 101.9 MHz is overpowering the channel. To solve the problem, you can build a tuning stub and connect it to the terminals of the FM receiver to trap out the local station. The antenna on the radio is 300 Ω, as is the impedance of the receiver. However, all you have to work with is some #24 wire, a ruler, thumb tacks, your Swiss Army knife, and your basic electromagnetics textbook.

Design a transmission line tuning stub to trap out the local station for your receiver. You must cut the local station by at least 10 dB relative to the clandestine station.

10.12

ANTENNA DESIGN

Suggested by
J. F. Corum
Battelle Memorial Institute

Universal Robotics, Inc., plans to develop a line of household service robots. Their first entry into the market will be a lawn-care robot that can keep the grass clipped on any estate up to 2 acres. To do this, the robot will use the Global Positioning System, GPS, to determine its position on Earth's surface to within 1/2 m. The GPS transmits two frequencies, 1575.42 and 1227.26 MHz, both of which are right-hand circularly polarized.

With the goals of maximizing gain and minimizing size, design one or more antennas to receive these signals. The design criterion is to maximize the ratio of gain to physical volume. An antenna should be omni-directional in the upper half plane. The antenna should have a gain of at least -2dB at the zenith and -3dB at 30° above the horizon. Additional information on the GPS is in the October 1983 issue of the Proceedings of the IEEE, vol 71, no. 10.

10.13

STEP-TO-PULSE FUNCTION CONVERTER AND DIFFERENTIATOR

Transmission Line Reflection Concepts

Submitted by
David T. Stephenson
Iowa State University

Design a passive transmission-line device that will convert a step function into a short pulse and that can also serve as a differentiator over a certain range of input signal parameters.

As the following figure 1 shows, the device is to be connected between a step generator and a 72-Ω resistive load.

The generator delivers (to a matched load) the step function shown in figure 2(a). (The generator output returns to zero after about one-half millisecond and the process repeats 1000 times a second, but only the step risetime is of interest to us here.) The device is to produce, across the load, a voltage pulse as shown in figure 2(b). Each pulse should have the same risetime characteristics as the step function, although it may have a reduced amplitude (V_{m2} is to be at least 20% of V_{m1}). The delay Δ through the device is not to exceed 400 ps. Subsequent spurious pulses or step functions at the output are to be no greater than 5% of V_{m2}.

Figure 3 shows a device made of transmission line segments that will perform this function. Specify the line lengths L_1 and L_3 and the values and tolerances for L_2 and the three characteristic resistances R_{01}, R_{02}, and R_{03}. Assume that the velocity of propagation in the line segments is $3 \cdot 10^8$ m/s.

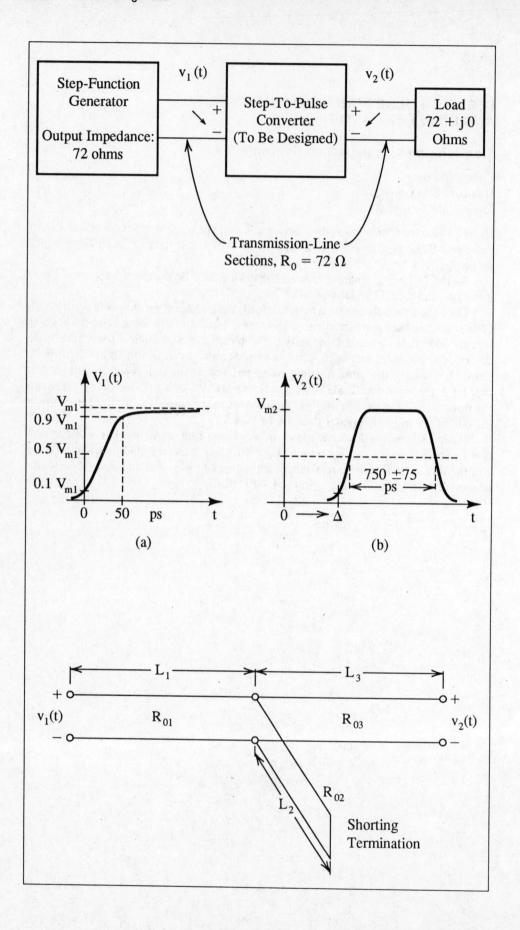

(a)

(b)

10.14

RADIO WAVE PROPAGATION AND ANTENNAS

Submitted by
T. Rappaport
Virginia Tech

In Washington DC, there is a building ordinance prohibiting structures taller than 170 m, which is the height of the Washington Monument. You are to design a 100-kHz LORAN-C navigation station in the city. The FCC requires you to prepare a detailed proposal of the transmitting antenna system, including number of radials, antenna efficiency, antenna impedance, and the received E-field at the following locations: New York City; Blacksburg, VA; Boston, MA; and Bermuda Island.

Determine the input power into the antenna matching system in order to provide a SNR of 20 dB to each of the four locations for both day and night operation. For noise purposes, assume Bermuda Island is a residential area and Blacksburg is rural. Your design should include losses experienced at the antenna base due to imperfect ground screen (assume the antenna matching unit is lossless), which cause the radiated power to be significantly less than the transmitter output power. Ground wave propagation is assumed and receivers use a 1-kHz bandwidth.

11

Power and Machines

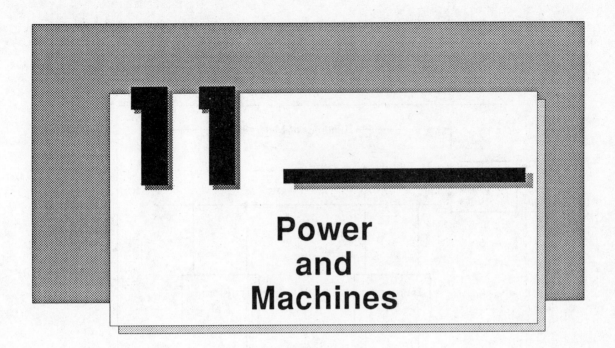

11.01

PUBLIC ADDRESS SYSTEM

A public address system—for example, at a hospital or high school—provides an audio signal to many speakers that are widely separated. Most speakers have approximately $8\ \Omega$ impedance and, for convenience, are connected in parallel. If all these speakers were to be connected directly to the PA amplifier, such a heavy current would be drawn that large wire would have to be used to carry it. Consequently, most systems use audio transformers to reduce current flow. These are called line-matching transformers and are designed to convert a 70.7 V rms audio signal to an appropriate level for an $8\text{-}\Omega$ speaker.

Design a transformer (choose the number of turns in the primary and the number in the secondary) to deliver 3.75 W of audio power to an $8\text{-}\Omega$ speaker. Neither winding should have less than 100 turns.

11.01 (continued)

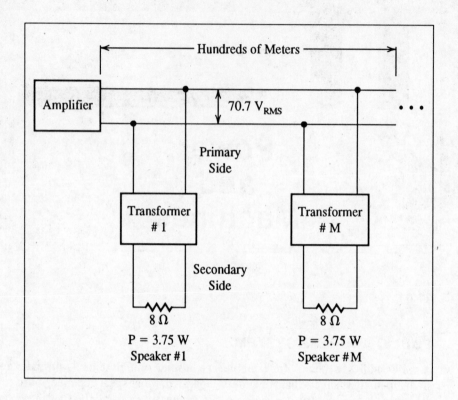

11.02

SOLENOID DESIGN

Submitted by
E. K. Stanek
University of Missouri-Rolla

Use the steel shell and plunger shown in the following figure to design a solenoid that will exert a net force of 130 lb on the plunger, including the effects of gravity, when the coil current is 5 A, the air gap, g, is 0.2 in., and the plunger weighs 5 lb. The magnetization curve for the iron structure is as follows:

| F, ampere-turns | 25 | 50 | 125 | 250 | 500 | 750 |
| ϕ, kilolines | 28 | 86 | 110 | 131 | 155 | 163 |

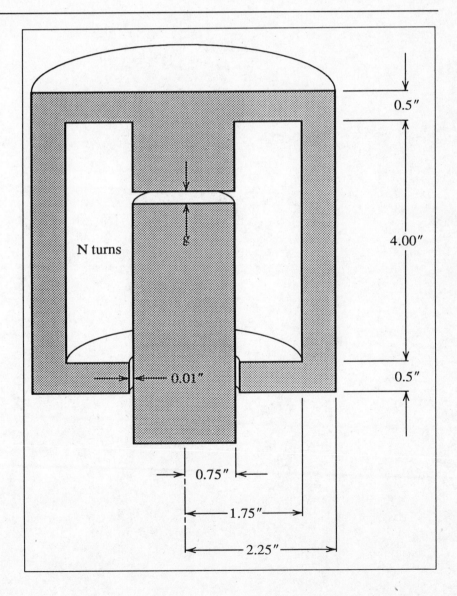

11.02 (continued)

Select copper conductors from the wire tables and assume a current density of 4 A/mm^2 is allowable because the current is not continuous. A packing density of 0.90 can be achieved with the conductors. (*Note:* The design of the coil entails the determination of the conductor size and the number of turns.)

11.03

ELECTROMAGNETIC DOOR DAMPER

You are to design an electromagnetic door damper to prevent open doors from being slammed shut. A coil of wire can be mounted on the door and a powerful cylindrical magnet mounted on a bracket so that the magnet enters the coil as the door closes. As the door closes, the movement of the coil induces a current and energy is dissipated in the resistor connected to the coil. The magnet is 20 cm long and 2 cm in diameter and is rated at 1500 Gauss.

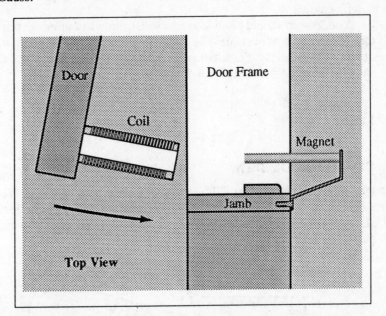

Design the coil and the resistor so that a typical door can be properly decelerated. Assume the door can be represented by a concentrated weight of 10 kg moving at a speed of 30 m/sec when the magnet first enters the coil. The door speed should be reduced to 2 m/sec by the time the door hits the jamb.

11.04

ARMATURE WINDING

Design a dc motor to work from an automobile battery, 12 to 14 V. The motor is to have four poles and produce a continuous 1 hp. The armature copper loss should not cause the armature temperature to rise more than 50°C above ambient; the thermal resistance of the armature is estimated to be 2°C for each watt dissipated. The armature dimensions are 4 cm diameter and 16 cm long. Because of magnetic field intensity constraints, no more than 50% of the surface of the armature may be removed for winding slots.

Your design should provide as many distributed windings as possible to make the torque as smooth as possible. The armature winding must be made with a single piece of wire, that is, no parallel windings or splices to change wire sizes. Draw your design showing how the windings are to be made.

11.05

DC MOTOR DESIGN

Design a small motor that will produce at least 5 N-m of torque. You have the basic magnetic structure as shown in the following figure, but you must decide how small the air gap (g) must be, how many turns of wire are required, and what current must be passed through the coil.

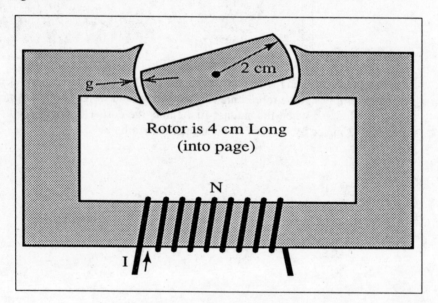

11.06

IMPACT OF PHYSICAL DIMENSIONS
ON DC MOTOR EFFICIENCY

Submitted by
E. K. Stanek
University of Missouri-Rolla

A 125-volt, 5-hp dc motor with conventional design has an efficiency of 80% at full load. Call this design C. Design a new motor with the same voltage and horsepower ratings and of the same general design as C but with a full load efficiency of 88.8%. Call this design D. Remember, efficiency is

$$\eta = \frac{P_{out}}{P_{out} + P_{losses}}, \text{ and that losses are related to physical size.}$$

In both designs, use the same material for windings, core, and insulation. The predominant losses occur in the copper and the core. Assume all other losses are negligible.

Design C has a diameter of 30 cm and an axial length of 50 cm. To keep designs C and D consistent, the length/diameter ratio should be maintained in design D. Assume both designs have the same number of coil sides and that the core material used results in hysteresis losses proportional to B^2.

Find the diameter and length of design D. Also indicate the relative weights and estimated relative costs of design C and design D.

11.07

ELECTROMAGNETIC GUN

Your employer is developing a "gun" to use in an explosive atmosphere, so exothermic reactions can't be used to propel a projectile. They have marketed a CO_2 gun, but want to explore another product. The company president, having seen a 60-Hz operated device that would throw off a large aluminum ring placed over it, asks you to design a gun to shoot a 1-g aluminum projectile horizontally with a muzzle velocity of 200 m/sec.

11.08

AUTOMATIC MOTOR STARTER

Your employer mines gypsum in a remote location. The company generates its own electrical power at 250 V dc. The gypsum is brought out of the mine on a long belt powered by a 500-rpm 20-hp shunt wound dc motor that has a full-load armature current of 75 A. This motor can't be started directly across the 250-V line because the mine power system can't supply more than 150 A to the motor.

Design a series of resistors with shunt contactors that can be used to start this motor under load (torque must not fall below rated torque). The contactors are to be operated automatically by voltage-sensitive relays. Draw the circuitry associated with the armature circuit and specify the resistance of each resistor used and the voltage threshold for each voltage sensitive relay.

11.09

DESIGN OF A RESISTANCE STARTER FOR A DC MOTOR

Submitted by
E. K. Stanek
University of Missouri-Rolla

A 20-hp dc motor is to be started from a 250-V dc supply. The motor has a full speed of 500 rpm, total armature resistance of 0.105 Ω, and a full-load armature current of 75 A. The field flux is at normal value during start up.

Design a multi-resistor starter that will meet the following specifications:

- Start-up current should not exceed twice the rated armature current at any time.
- Each time the starting current falls to rated armature current, a resistor should be switched out of the circuit.
- The minimum number of resistors should be used in the starter.

Include in your design the number of resistors needed, the ohmic value of each resistor, and the wattage rating of each resistor. You don't need to determine the length of the duty cycle on the resistors and you can neglect armature reaction and armature inductance.

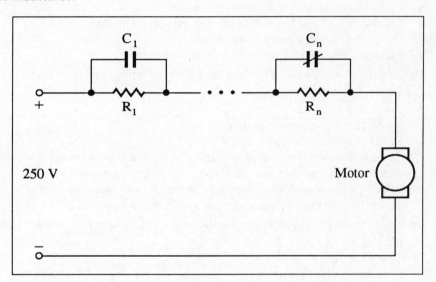

11.10

TRANSFORMER DESIGN BASED ON LINEAR DIMENSIONS AND VOLUME

Submitted by
E. K. Stanek
University of Missouri-Rolla

Determine the feasibility of designing a 200-kVA transformer based on an existing 25-kVA transformer design by making all linear dimensions of the 200-kVA transformer twice as large as in the 25-kVA one. Also investigate the impact that the change in dimensions would have on all the transformer ratings and parameters. Assume the following changes are made:

- All linear dimensions of the magnetic core are increased by a factor of 2.
- Lamination thickness is increased by a factor of 2 so that the number of laminations doesn't change.
- The magnetic core material is unchanged.
- The winding conductor diameters are doubled.
- The number of turns in each winding is unchanged.
- The winding insulation thickness is doubled.
- Conductor material and insulation material remain unchanged.
- Conductor current density is unchanged.
- Electrical stress on insulation in V/m is unchanged.

The original 25-kVA transformer had the following ratings and test values:

kVA rating = 25
Frequency = 60 Hz
Rated primary voltage = 2400 V
Rated secondary voltage = 240 V
Secondary current with rated secondary voltage and primary open = 1.066 A
Power in with rated secondary voltage and primary open = 126.6 W
Primary voltage with rated primary current and secondary shorted = 57.5 V
Power in with rated primary current and secondary shorted = 284 W

Determine the following quantities for the original transformer and the new design:

kVA rating
Copper losses
Core losses
Voltage ratings
Current ratings
Winding resistance in ohms and per unit
Leakage reactance in ohms and per unit
Exciting inductance in henries
Core loss resistance in ohms
Turns ratio

Does this approach produce an acceptable design for the 200-kVA transformer?

11.11

SELECTION OF LAMINATION THICKNESS TO CONTROL EDDY CURRENT LOSSES

Submitted by
E. K. Stanek
University of Missouri-Rolla

Design a single-phase transformer with the following ratings: 440 V: 220 V, 60 Hz, 5 kVA. The maximum allowable volume of the magnetic core material is 0.006 m^3. The core is made of steel with μ_r = 66,000 and ρ = 9.5 • 10^{-8} m-Ω. The number of primary turns is 200.

What are the acceptable values for core cross-sectional area and maximum lamination thickness if the eddy current losses are to be limited to 40 W? Assume B_{max} is not to exceed 1.0 tesla.

11.12

DOUBLE-CAGE INDUCTION MOTOR

Develop a double-cage rotor for a large motor to produce a high starting torque. The high starting torque is obtained by having a set of outer cage bars with a much higher resistance than the inner cage. At starting, the outer cage dominates because of skin effect. An approximate equivalent circuit for the rotor is shown in the following figure:

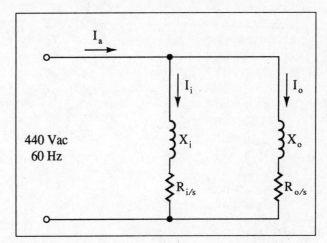

If X_i = 2 Ω and X_o = 1 Ω, choose values of R_i and R_o such that the outer cage produces 20 times as much torque as the inner cage does at starting but only one-tenth as much torque as the inner cage does at 2% slip.

11.13

INDUCTION MOTOR SPEED CONTROL
BY SELECTION OF ROTOR RESISTANCE

Submitted by
E. K. Stanek
University of Missouri-Rolla

A three-phase, wound rotor, wye-connected, 220-V (line-to-line), 20-hp, 6-pole induction motor has the following parameters in ohms per phase referred to the stator:

$R_1 = 0.294$
$R_2 = 0.144$
$X_1 = 0.503$
$X_2 = 0.209$
$X_0 = 13.25$

Assume the total friction, windage, and core losses are constant at 403 W, independent of load.

This motor drives a fan load that requires a torque of

$$T = 6.00 \cdot 10^3 \omega^2 \quad \text{N-m}$$

This load will be driven over speeds ranging from 1100 to 1150 rpm. The speed can be changed by adjusting the rotor resistance. If the ratio of stator turns to rotor turns is 3:1, determine the variable resistor to be added to the rotor windings to achieve the desired speed range. Your design should include the value of resistance and the wattage rating.

11.14

STARTING AND RUNNING CAPACITORS
FOR A TWO-VALUE CAPACITOR SINGLE-PHASE MOTOR

Submitted by
E. K. Stanek
University of Missouri-Rolla

A two-value capacitor single-phase motor has the following parameters:

Power rating:	1/4 hp
Voltage rating:	115V
Frequency:	60 Hz
Poles:	4

The ratio of turns in auxiliary windings to main winding is 1.10.

Main winding impedances:

$$r_1 = 2.20 \ \Omega$$
$$x_1 = 3.05 \ \Omega$$
$$x_m = 73.0 \ \Omega$$

Auxiliary winding impedances:

$$r_{1a} = 7.80 \ \Omega$$
$$x_{1a} = 3.52 \ \Omega$$

Rotor impedances referred to main winding:

$$r_2 = 4.50 \ \Omega$$
$$x_2 = 2.32 \ \Omega$$

No-load core loss: 23 W

No-load windage and friction: 15 W

Assume lossless capacitors are available for both starting and running capacitors.

Find the value of starting capacitor needed to produce auxiliary and main winding currents that are in quadrature. Also find the value of running capacitor needed to produce auxiliary and main winding currents that are in quadrature when the slip is 5%.

11.15

MATCHING A MOTOR AND LOAD

A commercial bread dough kneading machine that must be started when full will be driven by an electric motor that is not ideally suited to the job but which happens to be available. The motor will drive the mixer through a high-efficiency gear drive, and a suitable gear ratio needs to be determined.

With this kneading machine, the torque required decreases as the speed increases, as shown by the following tabulation of torque-speed characteristics:

Table 11.15-1. Kneading Machine Characteristics.

Speed (rpm)	Torque (1b/ft)
0	12.0
100	10.0
200	8.3
300	7.0
400	6.0
500	5.0
600	4.0
700	3.3
800	3.0
900	2.8
1000	2.6

11.15 (continued)

The machine will be driven by a single-phase ac motor rated 1/2 hp that starts as a repulsion motor and runs as an induction motor. The starting and running-torque characteristics are given as follows:

Table 11.15-2. Motor Characteristics
Repulsion Start-Induction Run
Rated 1/2 hp at 1750 rpm.

Speed (rpm)	Starting-Connection Torque (lb/ft)	Running-Connection Torque (lb/ft)
0	4.3	
200	4.7	
400	4.5	
600	3.8	0.9
800	3.1	1.3
1000	2.4	1.8
1200	2.0	2.4
1400	1.7	3.1
1600		3.7
1700		3.0
1750		1.5
1800		0

Transfer between the characteristics is accomplished automatically by means of a centrifugal switch mounted on the motor rotor, which operates at 1100 ± 50 rpm.

The power input to the kneading machine under constant load should be as great as possible within the limitations imposed by the available motor. What gear ratio do you recommend?

11.16

VOLTAGE SELECTION

Submitted by
Parman E. Reynolds, P.E.
Reynolds & Sterling

A new industrial plant has mostly motor loads that are served at 480 V. However, the plant load is geographically distributed so that a total of four substations are required to serve the load. The utility serving the plant will make either 4160 V or 13200 V available at the plant main switchgear. Based on the data, what would be the most cost effective design? Assume for this analysis that 1 hp = 1 kVA.

Load Data:

Sub #1:   Total 2000 hp, located 1000 circuit feet from main switchgear.

Sub #2:   Total 1000 hp, located 2000 circuit feet from main switchgear.

Sub #3:   Total 3000 hp, located 750 circuit feet from main switchgear.

Sub #4:   Total 850 hp, located 2500 circuit feet from main switchgear.

Equipment pricing:

Main switchgear costs:		4160 V system	13200 V system
	600 A sw	$ 8,000	$10,000
	1200 A sw	$10,000	$12,000
	1200 A c/b	$50,000	$60,000
	2000 A c/b	$75,000	$80,000

Bays can be fusible switch or circuit breaker type. A main switch will be required for either type of lineup.

Substation feeder costs per foot:	4160 V	13200V
100 A	$10	$15
200 A	$20	$23
300 A	$30	$35
400 A	$38	$43
500 A	$58	$65

Supplementary question:
    If the plant load were to increase in the future by approximately 50%, do you think your present voltage selection is correct?

11.17

INDUSTRIAL POWER SYSTEM DESIGN I

Submitted by
E. K. Stanek
University of Missouri-Rolla

A small shop has the following loads:

Motors:	50 hp	continuous
	10 hp	semicontinuous
	15 hp	semicontinuous
	25 hp	general purpose
	5 hp	general purpose
	5 hp	semicontinuous

Furnace (induction): 100 kW

Lighting:	75 kW	(90% ballast)
	50 kW	(90% ballast)

Arc Welder: 100 kVA

Using a growth factor of 50% (for all but lighting) and the attached demand factors, find the minimum ampacity of the service entrance. (*Note:* Normally no allowance is made for growth in lighting load.)

Complete the one-line diagram for the small shop shown in the following figure:

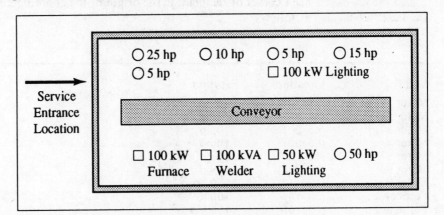

Include wire and conduit sizes (use THW cable), feeder breakers, fused switches, and combination fuse type motor control.

Use the ampacity found earlier to determine the plant demand in kVA and then select the service entrance and main breaker.

Typical demand factors:

Motors	
General purpose	30%
Semicontinuous	60%
Continuous	90%
Industrial Furnaces	80%
Lighting	100%
Arc Welders	30%

11.18

INDUSTRIAL POWER SYSTEM DESIGN II

Submitted by
E.K. Stanek
University of Missouri-Rolla

A manufacturing plant has the dimensions and loads positioned as shown in the following figure:

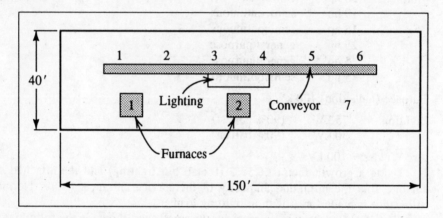

If the lower left-hand corner of the plant is the origin of the coordinate system, the loads are located as follows:

Load	Location	Rating
Motor 1	(25, 30)	10 hp
Motor 2	(45, 30)	10 hp
Motor 3	(65, 30)	10 hp
Motor 4	(85, 30)	10 hp
Motor 5	(105, 30)	10 hp
Motor 6	(125, 30)	10 hp
Motor 7	(120, 10)	150 hp
Furnace 1	(30, 10)	100 kW
Furnace 2	(75, 10)	100 kW
Lighting	(75, 20)	To be determined

11.18 (continued)

The plant is to have five feeders as follows:

Feeder #	Load
1	Motors 1 to 6
2	Lighting (90% power factor)
3	Motor 7
4	Furnace 1 (unity power factor)
5	Furnace 2 (unity power factor)

Assume the entire area is to be used for machine and rough bench work. Further assume the area will be illuminated with fluorescent fixtures to a level of 20 foot-candles, requiring 3.5 VA/ft^2.

11.18a.
The service entrance can be located anywhere on the building's outside walls. Pick the optimum location and justify it.

11.18b.
Size the conductors and conduits to supply the loads. Use type THW copper cable in nonmagnetic conduit and the following guidelines for feeder voltage drops. Allow for growth of 50% on all feeders except lighting feeders. Assume all motors are 80% efficient and have 0.85 power factors.
Guidelines for feed voltage drops:

Allowable drop on main circuits and feeders = 2%

Allowable drop on branch circuits = 1%

Remember, branch circuits should have at least 1/3 of the ampacity of the feeder circuit they are fed from. (Refer to NEC for voltage drop factors.)

11.19

RURAL DISTRIBUTION LINE EXTENSION

Submitted by
Parman E. Reynolds, P.E.
Reynolds & Sterling

A development company wants to build a large residential complex on its property. The load on the complex is estimated to vary from 50 to 500 kVA, depending on occupancy of the property. The nearest distribution line is a #2 ACSR single-phase line installed on 40-ft poles with 8-ft crossarms with the conductors set 6 in. in from each end of the crossarm. The line is energized at 13,200 V phase-to-phase. The existing line is approximately 15 mi from a utility substation and approximately 8 mi from the new residential complex.

As a distribution design engineer for the rural electric cooperative that will provide power for this project, you need to solve the following problems:

1. Will a #2 ACSR line extension provide adequate service in terms of voltage regulation? Estimated load swings are from 50 to 500 kVA at the complex.
2. You have investigated ground clearance and determined that a #2 ACSR center conductor could be installed on the pole line. Would this improve service?

You can find line electrical characteristics in a number of standard texts, including Westinghouse's *T & D* book and *Elements of Power System Analysis* by Stevenson*.

*Stevenson, W.D., Jr. *Elements of Power System Analysis,* 4th ed. McGraw-Hill Book Company. New York. 1982. 436pp.

Electrical Transmission and Distribution Reference Book, 4th ed. Westinghouse Electric Corporation. East Pittsburgh, PA. 1964. 824 pp.

11.20

STREET LIGHTING SYSTEM DESIGN

Submitted by
Parman E. Reynolds, P.E.
Reynolds & Sterling

A power company has a contract with the local municipality to provide design and construction of street lighting circuits for the main city streets. For a particular stretch of roadway approximately 4,800 ft long, the lighting designer has determined that the most effective system would incorporate lighting poles placed in the center of the street on 200-ft spacing. Each 40-ft tall pole would have two 250-W high-pressure sodium luminaries.

The overhead lines and poles available to serve the project are located in alleyways approximately 120 ft from the center of the street. The alleys are spaced at 400-ft intervals along and perpendicular to the street. It is contemplated that overhead transformer(s) will be installed in the alleyways and then underground cables run from the transformer to the street lighting system.

The following data apply to this situation:

- Transformers are available in standard single-phase kVA ratings from 10 to 167.5 kVA.
- A maximum of two transformer locations can be selected.
- Available transformer secondary voltages are 240 V or 480 V single phase.
- Height of the transformer off the ground is approximately 30 feet.
- System conductors will be 600-V copper conductors installed in nonmetallic conduit.
- Luminaire ballast operates at 0.90 lagging power factor.
- The internal pole wiring between base and luminaire will be #12 AWG serving both luminaries on the pole.

Design a street lighting system taking into account selected operating voltage, transformer location(s), transformer size(s), and conductor size. The maximum voltage drop to any luminaire should not exceed 5%. Neglect voltage drop through the transformer and burial depth of the underground circuits.

11.21

POWER SYSTEM LAYOUT

Submitted by
Paul Leiffer
LeTourneau University

Lay out in block-diagram form a system to generate power for a village of 30 persons in a remote area. A swiftly running stream is available within 200 m of the village. Specify the components as completely as possible. You will have to research components from sources other than the approved parts list.

11.22

BUILDING ELECTRICAL SYSTEM DESIGN

Submitted by
Parman E. Reynolds, P.E.
Reynolds & Sterling

A 40,000-ft^2 office building has the following electrical requirements:

Lighting load:	3.5 VA/ft^2, single phase
Air conditioning load:	240 kVA
Electric heating load:	400 kW
Miscellaneous load:	55 kVA
Receptacle load:	2 VA/ft^2, single phase

Using the *National Electrical Code* (NEC) as a guide, design the building service entrance equipment by calculating the following:

- Maximum connected load
- Minimum standard size of main circuit breaker
- Size of service entrance conductors
- Size of the feeder and feeder breaker to the air conditioning equipment

Neglect voltage drop in the feeder and service entrance cables. Assume the service voltage is 480 Y/277 V 3-phase, 4-wire and that copper conductors in steel conduits are used. Use the current edition of the NEC.

11.23

SINGLE PHASE DISTRIBUTION LINE

Submitted by
E.L. McConnell
Indiana Michigan Power Company

A power company is planning to build a single-phase power distribution line to provide service to the customers in the following table:

Table 11.23-1. Customer Data.

	X,Y Coordinates (mi)	kVA	pf (lagging)
A	0.5,0	100	.85
B	1,0	200	.9
C	1,1	100	.75
D	2,0	25	.75
E	2,1.5	75	.95
F	4,0	100	.80
G	6,0	50	.85
H	6,1	25	.95
I	7,0	25	.75

The peak demand shown for each customer is assumed to occur simultaneously. To allow for distribution losses, the primary voltage at the substation is regulated at 7500 V and must not drop more than 6% below this value at any customer. The cost to build a distribution line, excluding wire, is \$13,000/mi. You should use ACSR wire listed in the following wire table; the installed cost is \$10/lb.

11.23 (continued)

Table 11.23-2. Characteristics of Aluminum Cable Steel Reinforced (ACSR)Wire:

Size of Conductor (cir. mills or A.W.G.)	Weight (lb/mi)	Current Carrying Capacity (amperes)	Impedance (ohms/conductor/mi)
1,510,000	10,237	1340	$0.072 + j0.557$
954,000	6,479	1010	$0.113 + j0.585$
636,000	4,319	770	$0.169 + j0.609$
397,500	2,885	590	$0.259 + j0.636$
266,800	1,936	460	$0.385 + j0.660$
4/0	1,542	340	$0.592 + j0.776$
2/0N	970	270	$0.895 + j0.836$
1/0	769	230	$1.120 + j0.851$
2	484	180	$1.690 + j0.860$
4	304	140	$2.570 + j0.854$

Design this distribution line for minimum cost. Assume the substation is at the origin and the customer's location is given by the X, Y coordinates in the first table.

11.24

TRANSMISSION LINE DESIGN BASED ON SERIES INDUCTIVE REACTANCE

Submitted by
E. K. Stanek
University of Missouri-Rolla

Load flow and stability studies on an existing transmission system have indicated a new line is needed between two buses that are 150 mi apart. There are several factors to consider when designing a transmission line, including voltage drop, stability, corona, ampacity, etc.

The new line must have a series inductive reactance no greater than 0.040 per unit on a 1000 MVA base at 100 kV. It doesn't necessarily have to be rated at 100 kV; in fact, the voltage levels being considered include 138, 230, 345, 500, and 765 kV.

It has been decided that conductors no larger than 1033 MCM ACSR will be used. This conductor has a GMR of 0.0402 ft at 60 Hz. The option to use multiple conductors per phase (bundled conductors) should be considered. If two such conductors are used per phase, they will be located 1.5 ft from each other. If there are four conductors per phase, they will be located at the corners of a square 1.5 ft on a side. Because of corona problems, all 500-kV lines must have at least two conductors per phase and 765-kV lines must have at least four conductors per phase.

Following are the recommended minimal spacings between adjacent phase conductors at various voltage levels:

Voltage level (kV)	Spacing (ft)
138	15
230	20
345	30
500	36
765	45

11.24 (continued)

Assume single-circuit lines have equilateral spacing, while double circuit lines have super-bundle configuration shown as follows:

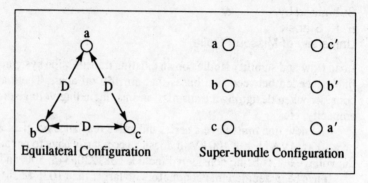

Equilateral Configuration   Super-bundle Configuration

Select the line design that will meet the design objective of 0.040 per unit series inductive reactance at 100.0 MVA, 100 kV and that will result in minimum conductor weight.

11.25

THREE-PHASE DISTRIBUTION LINE

Submitted by
E.L. McConnell
Indiana Michigan Power Company

A power company is planning to build a power distribution line to provide service to the following industrial and commercial customers:

Table 11.25. Customer Data.

	X, Y Coordinates	kVA	pf (lagging)	
A	1,0	500	.75	3 Φ
B	2,0	600	.80	3 Φ
C	2,2	250	.90	3 Φ
D	2,3	50	.85	1 Φ
E	2,4	50	.85	1 Φ
F	3,0	100	.80	1 Φ
G	3,1.5	350	.70	3 Φ
H	5,0	400	.80	3 Φ
I	7,0	200	.75	3 Φ
J	7,1	50	.90	1 Φ
K	9,0	50	.90	1 Φ

The peak demand shown for each customer is assumed to occur simultaneously. The minimum coincidental load is expected to be 30% of the peak load and the annualized average load is expected to be 55% of peak load.

The substation that will serve these customers is a nominal 12.5/7.2-kV grounded wye system but is regulated to hold 7500 V phase-to-neutral. The primary voltage must not drop more than 6% below this value at any customer. The projected load growth for these customers is 2%/yr and the distribution line is to be capable of adequately serving this load for 10 yr without modification, except for the possible addition of capacitors. Banks of capacitors or individual units can be installed if beneficial.

The grounded neutral on single-phase lines must match the primary wire size, and on three phase lines, it must have current carrying capacity of at least 65% of the primary wire. For the earth as a conductor in parallel with the grounded neutral, use an impedance of $0.286 + j3.25 \ \Omega/mi$.

The cost to build a distribution line, excluding wire, is $13,000/mi. ACSR wire listed in wire table in Exercise 11.23 is to be used; the installed cost is $10/lb. Single-phase capacitors rated 150 kVAR at 7.2 kV with an installed cost of $900/ea. can be used. The average cost of electricity at the substation is $0.0265/kWh. The carrying cost (interest plus maintenance cost) for this project is expected to be 24% of capital expenditure.

What is the most economical design for this line? Assuming the substation is at the origin, the customer's location is given by the X, Y coordinates in the table (in miles). The load is given in kVA, power factor and whether single phase or three phase.

11.26

TRANSMISSION LINE DESIGN

Submitted by
E. K. Stanek
University of Missouri-Rolla

Select a standard voltage level and design a transmission line that will supply a 250-MW load at 0.85 power factor lagging that is 200 mi from the source of power. The following specifications have been established for the line:

- The voltage regulation must be $\leq 10\%$.
- The efficiency must be $\geq 95\%$ at full load.
- One of the following standard voltage levels must be used: 138, 230, 345, 500, or 765 kV.
- A single circuit line with horizontal configuration is to be used.
- The following clearances between phase conductors must be maintained:

Line Voltage (kV)	Clearance (ft)
138	15
230	20
345	30
500	36
765	45

- To control corona, the following are the minimum size conductors or bundles of conductors that can be used.

Line Voltage (kV)	Conductor/Bundle	
138	500	MCM ACSR
230	715.5	MCM ACSR
345	2-1033.5	MCM ACSR (18" spacing)
500	2-1510.5	MCM ACSR (18" spacing)
765	4-1033.5	MCM ACSR (18" spacing)

11.27

ECONOMICS OF POWER FACTOR CORRECTION CAPACITORS

Submitted by
E. K. Stanek
University of Missouri-Rolla

A manufacturing plant has a variety of three-phase induction motors, furnaces, and lighting loads. It buys power/energy from an electric utility that has a two-part billing schedule as follows:

- Each kWh of energy costs $0.065 (energy charge).
- A demand charge of $5.00/mo for each kVA of peak demand averaged over a 30-min period is also charged.

The load of the plant is as in the following figure:

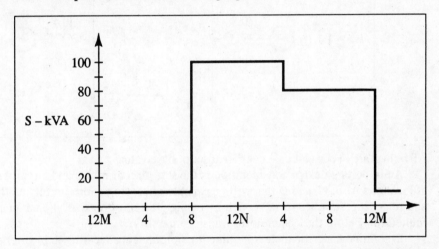

The load from 12:00 A.M. to 8 A.M. is largely lighting and ventilation. The power factor is 0.95 lagging. The load from 8:00 A.M. to 4:00 P.M. is the full load of the regular shift, during which everything is running, including several large furnaces making the power factor to 0.90 lagging. From 4:00 P.M. to 12:00 A.M., assembly and punch presses are run so that the furnaces used to heat-treat parts are shut down. The load consists of lightly loaded motors and a small percentage of lighting load. The net power factor is 0.80 lagging.

The manufacturing company is considering the purchase of power factor correction capacitors. These can be purchased for $50/kVAR in standard three-phase banks rated 10, 25, 50, 75, and 100 kVAR at 480 V. Recommend the optimum size bank of capacitors to purchase. Assume an interest rate of 12%/yr, or 1%/mo. Also, assume the capacitors will not be switched, but the utility charges no penalty for a leading power factor.

11.28

CRITICAL CLEARING TIME TO MAINTAIN TRANSIENT STABILITY

Submitted by
E. K. Stanek
University of Missouri-Rolla

A power system can be modeled adequately by the following two-machine equivalent circuit during a major system disturbance caused by a short circuit on one of the two parallel transmission lines:

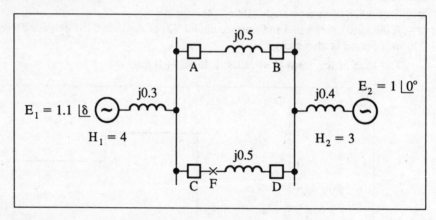

All values are in per unit on a consistent set of base values.

An important design consideration is the selection of relays and circuit breakers at locations A, B, C, and D that will be capable of sensing short circuits on the lines and removing the lines from service by opening circuit breakers at each end of the faulted line before the system loses transient stability.

Determine the critical clearing time for simultaneous clearing at locations C and D for a fault F on line C-D adjacent to breaker C to an accuracy of ±0.05 sec. The power transmitted from generator 1 to generator 2 prior to the short circuit was 0.4/unit.

In addition, determine if a short circuit located at any other point on transmission line C-D would lead to a smaller critical clearing time.

Based on the critical clearing time determined for this system, select the relay type and setting for locations A, B, C, and D. Also specify a circuit breaker speed in cycles based on 60 Hz.

11.29

CONNECTION OF A HARMONIC PRODUCING LOAD TO A UTILITY SYSTEM

Submitted by
Michael D. Higgins
American Electric Power

A prospective customer is planning to locate a rolling mill in your service area. The customer intends to install several 6-pulse dc drives at the plant. Because dc drives produce harmonics, you have been asked to determine if the resultant harmonic profile of the plant will be within your harmonic distortion guideline. If the harmonics would be in excess of the guideline levels, you are to recommend appropriate corrective measures.

The plant is to be connected to your system via a 138/13-kV, 20-MVA transformer with an impedance of 14% (on its own base). The available three-phase short circuit at the 138-kV point of connection is 1700 MVA at a 90% power factor with the 6-pulse dc drive load constituting about 70% of total load. The harmonic distortion limits for your utility at 138-kV are 5% total harmonic current distortion and 1% total harmonic voltage distortion. The harmonic profile of a typical dc drive is as follows:

Characteristic Harmonics = h = n P ± 1, where

P = the number of pulses

n = 1, 2. 3, etc.

Harmonic Current = $I_h = \dfrac{I_{60}}{h}$

h = harmonic order
I_{60} = magnitude of the 60-hz current consumed by the dc drive

Harmonic Distribution =

$$\frac{\sqrt{\sum_{h \geq 2} I_h^2}}{I_{60}} \quad \text{or} \quad \frac{\sqrt{\sum_{h \geq 2} V_h^2}}{V_{60}}$$

11.29 (continued)

Also, develop a simple computer program to calculate the currents of interest in the following harmonic filter circuits:

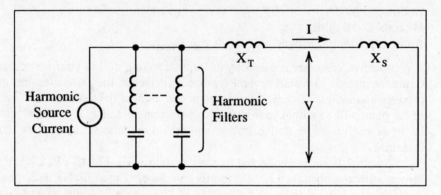

X_T = Transformer impedance
X_S = System impedance
I, V = Current, voltage at interface with the utility

11.30

EVALUATE SYSTEM PERFORMANCE

Submitted by
Richard J. Gursky
American Electric Power

The following figures 1 and 2 show a one-line diagram and a topological diagram for a model power system.

Evaluate the performance of this system against the given criteria and devise alternatives if the criteria are not met. Any alternatives should be evaluated from both technical and economic viewpoints. You should use a computer-based load flow package for the analysis and evaluate at least two alternatives. Alternatives should consider facility additions (for example, line, or capacitor) and/or reconfigurations of existing facilities. Use the following criteria:

- System should maintain acceptable performance for at least 5 yr.
- Area loads are growing at 5%/yr.
- Thermal loadings should be within "normal ratings" for noncontingency conditions.
- Thermal loadings should be within "emergency ratings" for contingency conditions.
- The system should be evaluated for only single contingencies, that is, outages of a single element at a time.
- Voltage drops should be less than 5% during a contingency.
- Voltage levels should not be less than 95% during normal conditions.
- The economic analysis should use a yearly interest rate of 12% for present worth calculations.

		Ratings	
Line	100 MVA base	Normal	Emergency
A-B	$0.8 + j\,4.0$	30 MVA	35 MVA
A-C	$0.4 + j\,2.0$	30 MVA	35 MVA
A-D	$0.2 + j\,1.2$	50 MVA	57 MVA
B-E	$1.0 + j\,5.0$	30 MVA	35 MVA
C-D	$0.4 + j\,2.2$	40 MVA	46 MVA
D-E	$1.6 + j\,8.0$	30 MVA	35 MVA

Facility Costs:	Lines	$200,000/mi
	Capacitors	$6/kVAR
Typical Facility Impedance:	$0.25 + j\,1.25\%$/mi (100 MVA base)	
Typical Facility Capabilities:	30 MVA (normal), 35 MVA (emergency)	

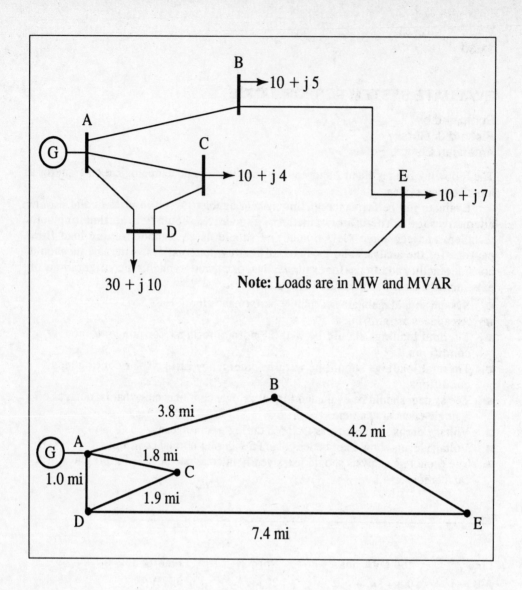

Note: Loads are in MW and MVAR

11.31

ECONOMIC CHOICE OF CONDUCTOR SIZE

Submitted by
Richard J. Gursky
American Electric Power

A system study has indicated the need for a new transmission line between two 138-kV stations (named A and B) that are 10 mi apart. The conductor for this new line must be sized with consideration to both thermal loading and costs, and should have enough capacity to handle both normal and contingency loadings during its lifetime.

After determining the minimum size of conductor (due to loading considerations), decide on the most economic size based on construction costs and the cost of losses. Because of reliability considerations, an additional circuit will be needed at some time between stations A and B.

The analysis should consider two means of construction: either single circuit for both lines or double circuit for the initial line with only one side strung.

Design the line assuming the following parameters:

- During normal conditions in the year of installation, the per unit voltages at stations A and B will be 1.05 $\angle 160°$ and 1.02 $\angle 19°$, respectively.
- During contingencies, the line loads to 150% of its normal loading.
- System loads are growing at 2%/yr.
- The economic analysis will be a 20-yr life.
- A parallel line will be installed in year 10 of the analysis.
- An inflation rate should be chosen based on current conditions.
- A carrying charge rate for capital investment is 20%/yr.
- A 12% interest rate applies to present worth calculations.

12 Communications

12.01

PROBABILITY

You negotiate a deal with a casino to design a new game. The game can use either ordinary dice or ordinary playing cards, should allow several persons to bet on the results, and should provide an expected payoff to the customers of 60 to 90¢ for each dollar bet so that the "house" can make a profit.

Devise such a game. Write the rules and compute the expected payoff for the customers.

12.02

PROBABILITY USING RESISTORS

The manufacturer of a particular printed circuit board encounters a crisis when the company runs out of 100-kΩ 5% resistors that are placed at a particular location on the board. It has plenty of 33-, 100-, 180-, and 220-kΩ 10% resistors. Each of these resistor values is uniformly distributed about the range within ±10% of its nominal value.

The company can use the 100-kΩ 10% resistors, three 33-kΩ 10% resistors in series, or the 180- and 220-kΩ 10% resistors in parallel. Whatever they do, if the actual resistance is not within ±8% of 100 kΩ, the board will not meet its performance specs.

Advise the company how to maximize the yield of acceptable boards and predict the yield under these conditions.

12.03

PSYCHOLOGY EXPERIMENT

As part of an experiment, a psychologist needs a way to choose a number at random between 1 and 10. The method must provide equal probability of selection for all 10 numbers. The psychologist has only one ordinary, six-sided die to use.

Devise a way to use this single die to randomly choose the desired numbers and provide written instructions for the procedure.

12.04

ERROR PROBABILITY OF A COMMUNICATION SYSTEM

You need a system to send four equally probable messages—m_0, m_1, m_2, and m_3. These are to be coded as a series of 0's and 1's. The probability that any particular 1 or 0 is received correctly is 90%. You know you can increase the probability of a correct transmission by increasing the length of the code word, but bandwidth limitations of the system make it impossible to use a word longer than 3 bits; consequently, there are eight such words (000, 001, 010, 011, 100, 101, 110, 111). Because there are only four messages to be coded, you have a choice in assigning code words to the four messages.

Devise an assignment scheme and a decision scheme at the receiver to minimize the error probability of the communication system. Your predicted performance should show the error probability of the system.

12.05

COMPACT CODE DESIGN

Morse code, invented by Samuel F.B. Morse (1791–1872) to be used with American telegraph systems, is still in use today. Unlike many codes you might be familiar with (such as ASCII), the Morse Code consists of elements of unequal length. The short pulse is called a dot (often sounded dit) and the long pulse is a dash (sounded dah). In the early telegraph, the dash lasted about three times longer than the dot, with a pause between code elements lasting about as long as one dot. The pause between English letters was as long as a dash. This means that a *d* would take about 10 units of time to transmit (dash, pause, dot, pause, dot, pause, pause, pause) whereas a *t* would take about six units of time.

Refer to the following table and use the probabilities of occurrence of the English letters shown as well as the Morse Code to explore the idea that Morse Code is a *compact code;* that is, the code words have been selected such that the length of time necessary to transmit an arbitrary message is shorter than the time required using some other code assignment. Can you make a more compact code than Morse did? Estimate how much more efficient your code is than Morse Code for sending English text. Ignore punctuation.

Morse Code and probability of occurrence of each letter.

A	didah	• −	0.0642	N	dahdit	− •	0.0574
B	dahdididit	− • • •	0.0127	O	dahdahdah	− − −	0.0632
C	dahdidahdit	− • − •	0.0218	P	didahdahdit	• − − •	0.0152
D	dahdidit	− • •	0.0317	Q	dahdahdidah	− − • −	0.0008
E	dit	•	0.1031	R	didahdit	• − •	0.0484
F	dididahdit	• • − •	0.0208	S	dididit	• • •	0.0514
G	dahdahdit	− − •	0.0152	T	dah	−	0.0796
H	didididit	• • • •	0.0467	U	dididah	• • −	0.0228
I	didit	• •	0.0575	V	didididah	• • • −	0.0083
J	didahdahdah	• − − −	0.0008	W	didahdah	• − −	0.0175
K	dahdidah	− • −	0.0049	X	dahdididah	− • • −	0.0013
L	didahdidit	• − • •	0.0321	Y	dahdidahdah	− • − −	0.0164
M	dahdah	− −	0.0198	Z	dahdahdidit	− − • •	0.0005
Space			0.1859				

*Abramison, N. Information Theory and Coding, McGraw-Hill Book Co. New York. 1963. p 34.

12.06

SATELLITE COMMUNICATIONS ENHANCER

Your employer is designing a weather reporting satellite. The observing system is steerable so that as the satellite passes overhead, a relatively small portion of Earth can be studied for extended periods. Steering is accomplished by a communication system that sends two unique, equally probable messages as a series of 1's and 0's. The system requires that the messages be interpreted correctly 99% of the time; unfortunately, the communication channel is noisy, so there is only a 90% probability that any particular 1 or 0 is received correctly. The system's accuracy can be improved by coding the two messages as long strings of 1's and 0's so that the received message is interpreted correctly in spite of transmission errors.

Design such a system by specifying patterns of 1's and 0's that can be sent representing each of the two messages and that can be interpreted with 99% accuracy. These strings of data should be no longer than necessary to accomplish the objectives.

12.07

ORTHONORMAL SERIES DESIGN

Often, you might want to use a finite orthonormal series to represent certain signals over an interval. With this in mind, design (choose all the coefficients for) an orthonormal series consisting of the four simplest polynomials of t over the interval $0 \le t \le 2$.

$$y_0 = a_{00}$$
$$y_1 = a_{11}t + a_{10}$$
$$y_2 = a_{22}t^2 + a_{21}t + a_{20}$$
$$y_3 = a_{33}t^3 + a_{32}t^2 + a_{31}t + a_{30}$$

12.08

ORTHONORMAL POLYNOMIALS

A 3-year-old prodigy has reproduced the Taj Mahal and the Kremlin using her building block sets. Design a new building set to challenge her—an orthonormal set of functions for her to use to build images on her computer. You elect to produce a set of polynomials in x.

The polynomials should look like

$$u_0 = \alpha_{00}$$

$$u_1 = \alpha_{11}x + \alpha_{10}$$

$$u_2 = \alpha_{22}x^2 + \alpha_{21}x + \alpha_{20}$$

$$u_3 = \alpha_{33}x^3 + \alpha_{32}x^2 + \alpha_{31}x + \alpha_{30},$$

etc.

which are orthonormal over the interval $0 \le x \le 1$.

Provide only enough polynomials to approximate a rectangular function to within 5% rms error. Design the polynomial set that meets the criteria for the test function shown in the following figure:

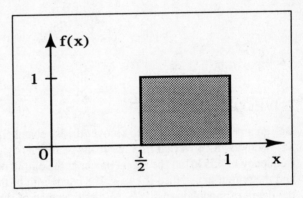

12.09

WALKIE-TALKIE DEMODULATOR DESIGN

Your employer builds low-cost walkie-talkie toys. The units have been designed to transmit a 1-MHz amplitude-modulated audio signal.

Design a demodulator whose output is to go to an amplifier with a 1-kΩ resistive input impedance. The source impedance R_s of the signal to be demodulated is 5 Ω. The walkie-talkie should have a frequency response of 500 to 2000 Hz. Determine the value of capacitor C that should be used in the design.

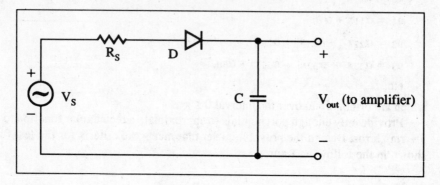

12.10

ENVELOPE DETECTOR

Design an envelope detector to recover the baseband signal from a full AM carrier that has been down-converted from a broadcast frequency of 1510 kHz to an intermediate frequency of 455 kHz. The signal power at the output of the 455 kHz IF filter is 0 dBm and the output impedance is 1 kΩ. The bandwidth of the baseband signal is 5 kHz and the depth of modulation is 90%. A circuit sketch of the envelope detector is shown in the following figure:

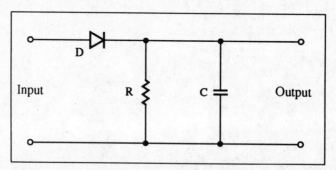

Specify the passive elements to be used in the first-order low-pass filter. Ensure the design fully recovers the baseband signal, while minimizing the output impedance of the detector. Assume the diode is ideal.

12.11

SINGLE-BALANCED MIXER

A single-balanced mixer is constructed from a 4016 CMOS analog switch that is modulated by a 2-MHz square-wave oscillator, producing a "chopped" or naturally sampled version of the input signal.

Given that a 50-Ω source generates an input signal with an rms value of 1 V and bandwidth of 15 kHz, design a filter that passes only the mixed signal at 2 MHz and suppresses (by at least 30 dB) all other spectral components appearing in the signal from the analog switch. Assume the 4016 acts as an ideal switch.

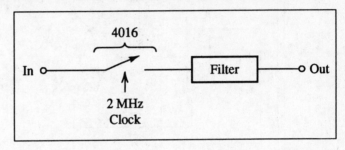

12.12

DOUBLE-BALANCED MIXER

The circuit sketch of a "ring diode" double-balanced mixer is shown in the following figure:

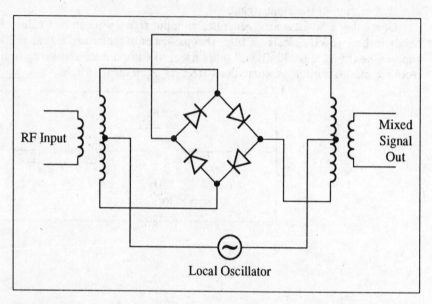

This balanced circuit prevents feedthrough of both the input signal and the local oscillator signal. The input signal power is −5 dBm and the local oscillator signal is approximately +5 dBm, which is large enough to cause the diodes to approximate ideal switch behavior.

Specify the turns ratios to be used in the transformers so that both the input and output impedances are 50 Ω at 10 MHz. Assume the input source impedance is 50 Ω and the output of the mixer feeds a 50-Ω load. Also, determine the impedance seen by the local oscillator.

12.13

UNBALANCED FM DISCRIMINATOR

An FM signal is down-converted in an FM receiver to a 455-kHz intermediate frequency carrier before being fed to an unbalanced discriminator for baseband recovery. The signal power at the output of the 455-kHz IF filter is 0 dBm and the bandwidth is 150 kHz. The base bandwidth is 15 kHz.

Specify the values of the components to be used in the FM discriminator circuit shown in the following figure:

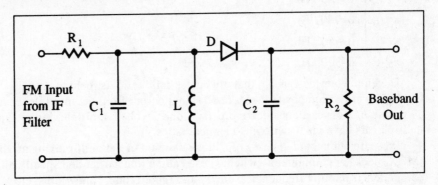

12.14

FREQUENCY MULTIPLIER

Frequency multipliers often are needed in communications systems to expand the spectrum of a signal. One method of doing this involves extracting a high harmonic from a limited (square-wave) representation of the input signal by tuning a resonant circuit to the frequency of that harmonic. The frequency of the extracted harmonic is then an integer multiple of the fundamental frequency of the original input signal.

Design a resonant circuit for a frequency tripler that has a 5-V, 335-kHz square-wave input, as shown in the following figure. The impedance of the square-wave source is $(10^5 + j0)\ \Omega$. The third harmonic should be greater than any other harmonic by at least 10 dB.

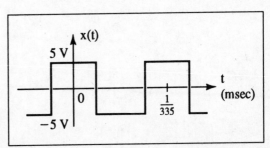

12.15

PAM-TDM TELEMETRY SYSTEM

Design a PAM-TDM telemetry system for communicating five analog signals and a synchronization marker. The analog signals are low-pass and are specified in terms of "absolute" bandwidths, implying that the signals have no spectral content outside of these bandwidths. The signals and their absolute bandwidths are as follows:

x_1 (t): 0–4500 Hz

x_2 (t): 0–1050 Hz

x_3 (t): 0–900 Hz

x_4 (t): 0–450 Hz

x_5 (t): 0–300 Hz

Develop your design so that the cross-talk ratio is below -30 dB and the bandwidth of the composite PAM-TDM signal output from the multiplexer is as low as possible. You can consider more than one level of multiplexing. Assume the multiplexers have ideal switching characteristics.

Present your design in terms of a block diagram of the multiplexing arrangement (MUX blocks or commutator switches) used at the transmitter and specify all system parameters (sampling frequencies, clock rates, guard time, bandwidths, etc.). Present a second block diagram of a receiver layout for the PAM-TDM system that shows signal demultiplexing and reconstruction.

12.16

PCM-TDM COMMUNICATIONS SYSTEM

Suggested by
Theodore Rappaport
Virginia Tech

Your job is to design a digital microwave radio communications system for a common carrier (the telephone company). The system uses OOK modulation and operates at a carrier frequency of 2 GHz. The link is to accommodate 24 simultaneous voice channels. These channels are individually sampled and then time-division multiplexed (TDM) as one baseband message, which is then applied to the input of a DSB-SC transmitter.

Develop your design by showing a complete block diagram of the transmitter and receiver, and generate a set of specifications covering the items listed below. Assume each voice channel has a maximum frequency component of 5 kHz.

1. the sampling rate for each voice channel.
2. the duration over which a single voice channel is sampled assuming uniform dwell time on the commutator.
3. the quantization SNR of the system.
4. the number of binary PCM bits used to encode each channel.
5. the data rate of the TDM-PCM binary data stream.
6. the minimum bandwidth required for the baseband data stream and the type of pulse shaping needed to achieve this bandwidth.
7. the parameters of a raised cosine-rolloff filter which precedes the DSB-SC transmitter which results in the RF bandwidth being limited to 3 MHz.

12.17

BINARY COMMUNICATIONS RECEIVING SYSTEM

Submitted by
Joseph H. Wujek
University of California at Berkeley

A binary communications receiving system examines an incoming (serial) bit stream, decodes the stream into bytes of eight-bit lengths, and transmits these bytes. The system can detect and correct bit errors provided no more than one bit error occurs in the eight sequential bits to be converted to a byte. The bit-error rate of incoming signals is 10^{-6} (one error per million bits).

Assuming a ground benign environment with ambient temperature $15 \leq T \leq 35°C$, design a circuit to perform the function described above. Use any logic family you want that is consistent with the design information given in this exercise.

DESIGN INFORMATION and SPECIFICATIONS

The failure rate of your design is not to exceed $1 \cdot 10^{-7}$/hr.

The bit-stream data rate is 5 MHz, with rise and fall times between 10 ns and 25 ns. The jitter in marking the start and end of each bit's time-interval is ± 5 ns.

All signals (including those in your design) have rise/fall times measured between 10 and 90% amplitude and pulse-width measured at 1.5 V amplitude. All signals (including those in your design) are high TRUE, 4.1 to 5.0 V amplitude for ONE and 0.0 to 0.4 V for logic ZERO.

Each eight-bit byte is to be stored in a register for parallel output, then transferred out in parallel in response to a DUMP DATA signal at 40 ± 1 ns after the parity bit is loaded into the register. This output function is to be completed ≤ 25 ns after the (leading) positive edge of the DUMP DATA signal that is 40 ns wide with edges ≤ 10 ns.

The serial data input is transmitted in 10-bit sequences. The bit in the first bit interval is not output in your circuit design. This is followed by seven bits of data, B0–B6, transmitted with the least significant bit (LSB) first and MSB last. Immediately following the MSB is the PARITY bit using the EVEN-PARITY convention. Then, immediately following the PARITY bit is another bit that is not output in your design. This input sequence is summarized as follows using colons to separate the 200 ns (nominal) bit intervals (BLANK indicates "not used"):

... :BLANK:(LSB) B0: B1: B2: B3: B4: B5: (MSB) B6: PARITY: BLANK: ...

The SYNC used for flagging the start of a byte is derived from another circuit and is furnished on a wire, distinct from the bit stream, for use in your design. The signal SYNC is the positive edge of a pulse 100 ± 5 ns in width. This positive-edge reaches 1.5 V no later than 10 ns after the interval associated with the LSB begins. At all other times, SYNC is at ZERO. The circuit that generates SYNC can source or sink current up to 1 mA, and your circuit's interface to SYNC should require 1 mA in magnitude.

Make any further assumptions you might need to develop your design, provided these don't conflict with the specifications or criteria given earlier in this exercise.

12.18

60 HZ DIGITAL NOTCH FILTER

Submitted by
Lang Tong
West Virginia University

The audio feed line to an RF modulator is too close to 60-Hz power conductors and the 60-Hz hum is reducing the quality of the audio portion of the broadcast signal.

Design a second-order IIR digital filter to remove the 60-Hz hum while not degrading the quality of voice signals any more than necessary. The audio bandwidth is 20 to 20,000 Hz. Also, draw a block diagram of your system, specifying the functional requirements of each block.

OPTIONAL: Draw a complete circuit diagram of your system using components from the approved parts list.

12.19

SPEAKING BACKWARDS

Digital Signal Processing

Submitted by
Lang Tong
West Virginia University

AcmeCom, Inc., is developing an encryption system to provide security for telephone conversations. The method of encryption involves simply separating the words in a sentence and then transmitting them in reverse order. A similar system is used at the receiving end to reverse the process. Assume the words are spoken distinctly so that the energy level drops substantially between words, and that there is a recongnizable pause at the end of each sentence.

Auxiliary circuits generate signals to indicate the end of words and the end of sentences.

Design the digital subsystem to reverse the speech. The bandwidth is 300 to 3000 Hz. Also, specify the requirements of each block in the block diagram of your system.

12.20

MALE TO FEMALE VOICE CONVERSION

Digital Signal Processing

Submitted by
Lang Tong
West Virginia University

A male voice has been digitally recorded into a memory. You are to produce a female voice from the recorded message without using sophisticated filters and processing techniques. Design the algorithms to do this.

12.21

LOW-PASS FILTER

Signal Processing

Submitted by
Lang Tong
West Virginia University

An analog signal $x(t)$ consists of the sum of two components, $x_1(t)$ and $x_2(t)$. The spectral characteristics of $x(t)$ are shown in the following figure. The signal $x(t)$ is bandlimited to 40 kHz.

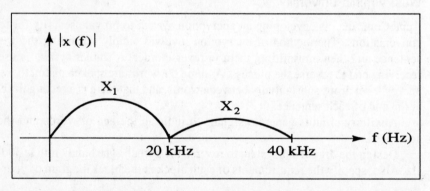

Design a Butterworth low-pass filter to suppress signal $x_2(t)$. The allowable amplitude distortion of $|X(f)|$ is ±2dB over the range of $0 \le |f| \le 15$ kHz. Above 20 kHz, the filter must have an attenuation of at least 40 dB.

12.21a.
Design an IIR digital filter using a bilinear transform to perform the function in 12.21.

12.21b.
Design an FIR filter using a Hamming Window to perform the function in 12.21.

12.22

FINITE IMPULSE RESPONSE FILTER

Digital Signal Processing

Submitted by
Alvin Moser
Seattle University

Design a finite impulse response (FIR) filter having the following amplitude response:

$$A_d(f) = 0 \quad \text{for } 0 \leq f < 60 \text{ Hz}$$
$$= 0.5 \quad \text{for } 60 \text{ Hz} \leq f < 100 \text{ Hz}$$
$$= 1 \quad \text{for } 100 \text{ Hz} \leq f$$

The sampling rate should be 1 kHz, and the impulse response should be 40 ms long. Using the following triangular window function,

$$w(m) = 1 - \frac{|m|}{20} \text{ for } |m| \leq 20,$$

design the transfer function H(z) and implement it as a program. Your documentation should show results of tests on the program for input frequencies of 0 to 150 Hz at 10-Hz intervals.

12.23

BILINEAR FILTER

Digital Signal Processing

Submitted by
Alvin Moser
Seattle University

Design a digital low-pass filter using the bilinear transformation. The sampling rate should be 10 kHz and the cut-off frequency near 1 kHz. Important to this design is good low-frequency correspondence. The prototype should be a third-order Chebyshev filter, with 1-dB ripple implemented as a program. Your documentation should include tests of the program for input frequencies from 0 to 1.5 kHz at 50-Hz intervals.

12.24

DIGITAL TONE GENERATOR

Digital Signal Processing

Submitted by
Alvin Moser
Seattle University

You need to digitally generate sinusoidal tones of limited duration. Rather than using extended lookup tables for the tones, your group leader has decided to try generating them algorithmically. You know that stability of transfer function H(z) is determined by the location of poles relative to the unit circle in the z-plane. As the poles approach the unit circle from the stable side, the impulse response takes longer and longer to stop ringing.

Devise a method of producing sinusoidal tones of period 16, 32, and 64 samples with "marginally stable" transfer functions. Investigate what performance improvements could be obtained with double-precision versus single-precision floating point coefficients.

12.25

TEST SCORE VALIDATION

Digital Signal Processing

Submitted by
Alvin Moser
Seattle University

A professor has been accused of arbitrarily lowering, over the last two years, the final score of every fourth student by 10 points when assigning grades. You, as a member of an academic grievance committee, have been asked by the committee's chairman to design a program to analyze the 256 scores assigned over the 2-yr period. You can make the following assumptions:

- True scores average 72 points.
- The student scores have a Gaussian distribution.
- The standard deviation of the scores is 12.
- Each score is independent of any others.

In summary, the true scores can be regarded as Gaussian white noise with a mean of 72 and a standard deviation of 12.

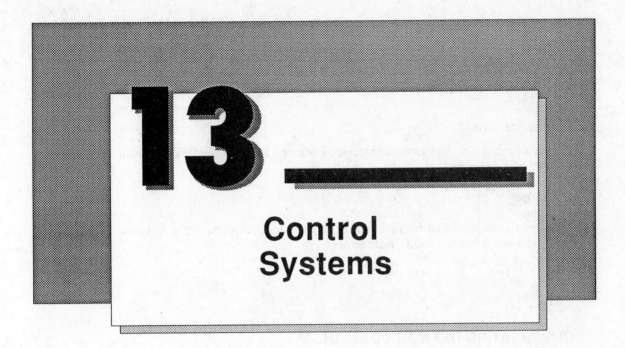

13

Control Systems

13.01

SYSTEM SENSITIVITY TO DISTURBANCE SIGNALS

Submitted by
David H. Thomas
West Virginia University

Your employer has a problem with one of its vertical positioning systems, largely because of the effect of gravity. The open-loop block diagram of the positioning system and the inaccessible disturbance (gravity) are shown in the following figure:

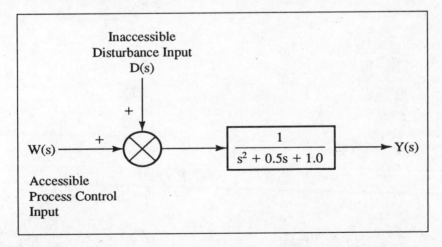

Design a way to decrease by 90% the effect of the disturbance on the positioning system by adding a controller block. Verify in your design that the new system is stable. Your predicted performance should include the steady-state error and the impulse response of the system.

341

13.02

ROTATING ANTENNA CONTROL SYSTEM

Submitted by
Paul Leiffer
LeTourneau University

A system is needed to control a rotating antenna. The motor has a transfer function given by the following equation:

$$T(s) = \frac{2.36}{(s + 20)}$$

Design (specify the transfer function for) a feedback system having a damping factor of 0.4 and minimum steady-state error.

13.03

COMPENSATING NETWORK DESIGN

Submitted by
Paul Leiffer
LeTourneau University

A compensating network is required whose transfer function is given by the following equation:

$$\frac{V_o}{V_i} = \frac{10\,(s + 2)}{(s + 3)}$$

Design a circuit that will yield this transfer function. Assume V_i is from a zero-impedance source and that V_o will drive into an infinite impedance.

13.04

LINEAR FEEDBACK CONTROLLER FOR A FIELD-CONTROLLED DC MOTOR

Submitted by
Craig S. Sims
West Virginia University

A model of a field-controlled dc motor is shown in the following figure:

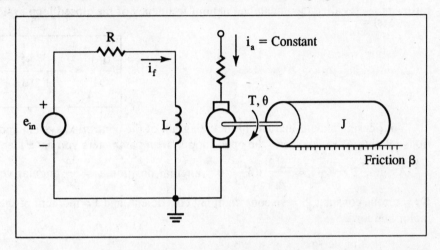

Draw a block diagram with known parameters to facilitate the design of a linear feedback controller. The block diagram should show the transfer function between input voltage and field current and that between field current and angular velocity. Design a process of identifying open-loop system parameters by putting in a sinusoidal input voltage and measuring field current and angular velocity. Also, design a speed controller based on the use of a linear combination of a command reference signal and feedback signals from a tachometer and current meter, that is

$$e_{in} = A [r - k_1 \omega - k_2 i_f].$$

13.04 (continued)

The specifications for the system are concerned with stability and speed of response, as well as steady-state error to a step change in command reference signal, r. Design your controller so that there is zero steady-state error to a step and settling time is about 1 sec. The controller must be stable. Percent overshoot for a step is used as a measure of relative stability; it has been determined that 15% is acceptable. The *settling time* is the time required for the speed to remain within about 5% of its final value. A common measure of settling time is $t_s = 3/\varsigma\,\omega_n$, where ς is called the damping ratio and ω_n is called the undamped natural frequency of the closed loop system,

$$\frac{\Omega(s)}{r(s)} = \frac{\omega_n^2}{s^2 + 2\zeta\omega_n s + \omega_n^2}$$

Your design should include a block diagram of the controlled system and equations relating A, k_1, and k_2 to the open-loop system parameters you have identified.

Assume $T = K_T\,i_f = J\dfrac{d^2\theta}{dt^2} + \beta\dfrac{d\theta}{dt}$, θ = angular position, $\omega = \dfrac{d\theta}{dt}$ angular velocity,

K_T = torque constant, β = viscous damping coefficient, and J = moment of inertia of motor and load.

13.05

ROBOT MANIPULATOR CONTROL

Submitted by
Gene Moriarity
San Jose State University

Design a feedback control system for a robot manipulator such that the steady-state position error $e_{ss} \leq 0.1$ when the input r(t) is a unt step. The transfer function of the positioning system is as follows:

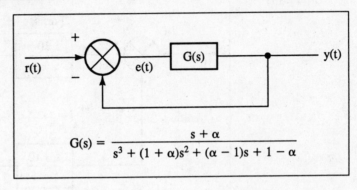

$$G(s) = \frac{s + \alpha}{s^3 + (1 + \alpha)s^2 + (\alpha - 1)s + 1 - \alpha}$$

13.06

ELECTRIC FURNACE CONTROL

Submitted by
Gene Moriarity
San Jose State University

You are working on a controller for an electric furnace used in a critical manufacturing process. The furnace and the controller have the following characteristics:

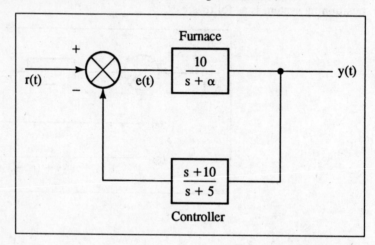

What value of the open-loop pole of the furnace would be best controlled by this controller, that is, that minimizes the integral squared error (ISE)? Assume the input is $r(t) = e^{-5t} u(t)$.

13.07

WATER LEVEL CONTROL

Controllability/Observability

You have been assigned to a design team that is designing a control system to maintain the liquid in a tank at a desired level. The team has already chosen to monitor the liquid level by using a float to determine the liquid depth, $h(t)$. The open-loop control input is defined as $e_i(t)$. The team has identified the system parameters and equations as follows:

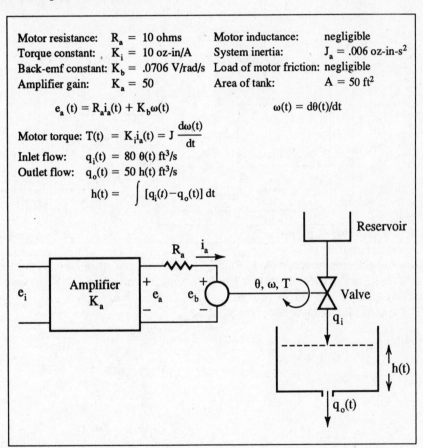

Motor resistance: $R_a = 10$ ohms   Motor inductance: negligible
Torque constant: $K_i = 10$ oz-in/A   System inertia: $J_a = .006$ oz-in-s^2
Back-emf constant: $K_b = .0706$ V/rad/s Load of motor friction: negligible
Amplifier gain: $K_a = 50$ Area of tank: $A = 50$ ft^2

$$e_a(t) = R_a i_a(t) + K_b \omega(t) \qquad\qquad \omega(t) = d\theta(t)/dt$$

Motor torque: $T(t) = K_i i_a(t) = J \dfrac{d\omega(t)}{dt}$

Inlet flow: $q_i(t) = 80\ \theta(t)$ ft^3/s
Outlet flow: $q_o(t) = 50\ h(t)$ ft^3/s

$$h(t) = \int [q_i(t) - q_o(t)]\ dt$$

Your job is to do the following:
- Determine if the open-loop system is completely controllable.
- For reasons of economy, only one variable may be fed back for the closed-loop control purposes. The team has identified three candidate variables, $h(t)$, $\theta(t)$, and $\omega(t)$. Recommend which variable should be used to produce a completely observable system.

13.08

FLUID LEVEL CONTROLLER

Feedback Controller

Referring to Exercise 13.07, it has been suggested that the loop could be closed as shown in the following figure:

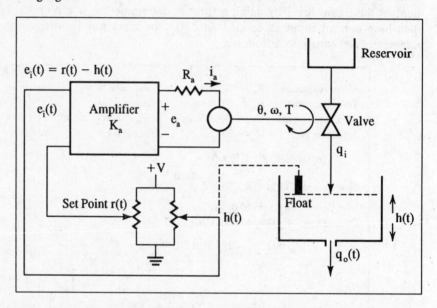

To make the system most economic, it must respond as quickly as possible while remaining asymptotically stable. The water control valve is available in several sizes, giving flow rates of 20, 40, 60, 80, and 100 cu ft/sec/rad (θ).

Choose the best valve for the system.

13.09

PID CONTROLLER

To minimize vibrational effects, a telescope is magnetically levitated. This method also eliminates friction in the azimuth magnetic drive system. The photodetectors for the sensing system require electrical connections, which can be represented by a spring with a constant of 1 kg/m. The mass of the telescope is 100 kg. The equation of motion for the azimuth drive is

$$f(t) - K_s y(t) = M \frac{d^2 y(t)}{dt^2}.$$

Design a PID controller for the system so that the ramp error constant $K_v = 100$ and the maximum overshoot is less than 5%.

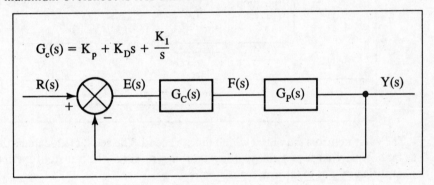

13.10

ACIDITY CONTROLLER

Acidity of water draining from coal mines is often controlled by adding enough lime to the water to make a neutral solution. A system to control the acidity is shown in the following figure:

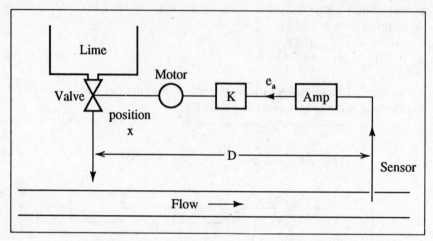

The valve controls the rate at which lime is added. The sensor is located a distance D as far downstream as possible to allow thorough mixing. The transfer function from the amplified output to the valve position is

$$\frac{x(s)}{e_a(s)} = \frac{K}{s^2 + 10_s + 100},$$

where e_a is the amplifier output in volts, K is a variable gain, and x is the valve position in revolutions. When neutral water is sensed, $e_a = 0$, when maximum acidity is sensed, $e_a = 10$ V; when maximum alkalinity is sensed, $e_a = -10$V. The valve can move between 0 (fully closed) and 1.0 (fully open). The stream flow rate can be assumed to be constant at 1 ft/s.

Design the system by choosing K and the distance D to maintain the stability.